STUDENT'S GUIDE AND WORKBOOK
TO ACCOMPANY

ELEMENTS OF
STATISTICAL REASONING

STUDENT'S GUIDE AND WORKBOOK TO ACCOMPANY

ELEMENTS OF STATISTICAL REASONING

EDWARD W. MINIUM
ROBERT B. CLARKE

prepared by

GORDON BEAR
Ramapo College of New Jersey

JOHN WILEY & SONS, INC.
New York • Chichester • Brisbane • Toronto • Singapore

ISBN 0 471 86463 3
Printed in the United States of America

10 9 8 7 6 5 4 3 2 1

Contents

Acknowledgments

Many people deserve thanks for their contributions to this book. I am grateful to Ed Minium and Bob Clarke for the inspiration they provided in the form of their unusually fine text *Elements of Statistical Reasoning*, to which this *Guide* is a companion, and for their valuable advice on this *Guide*; to Bob Abelson, Barry Collins, and Fred Sheffield for their conscientious instruction as my own teachers of statistics; to my teaching assistants at the University of Wisconsin, who worked diligently with me and provided helpful advice for improving my instruction; to my students at Wisconsin and at Ramapo College, from whom I have learned a great deal about pedagogy; and to my colleagues near and far Gary Beecher, Jane Goldman, Mike Lougee, Elsa and Les Wormack, and Bob Worsham, for welcome help with a variety of matters.

In spite of their many other obligations, the staff of John Wiley & Sons offered me all the personal attention I needed. I am especially indebted to Jack Burton, Elizabeth Doble, Carol Luitjens, Anita Kann, and Pam Cassell for their expertise and their good will.

I created this book via the computerized word-processing system SUPER-WYLBUR, and for help with the hardware, the software, and the operation thereof, I am grateful to the staff of the New Jersey Educational Computing Network, especially Dan Devine, Elaine Kaltenbach, Frank McArthur, Dave Norman, Eva Valentine, and Carol Yunker. At Ramapo College, Sue Wagner, Diane Coles and others in the computer center provided valuable assistance, and the College paid for my computing time. Chris Muller of Muller-Mugno Corporation and Ray Roesser of Science Typographers taught me everything I know about compositor codes, and I am particularly indebted to Daniel Breen of Anderson Jacobson, Inc. Dan supplied my hardware at a fair price, and it worked flawlessly: an AJ 880 teleprinter, an AJ 460 Micro Diskette System for offline storage, and an AJ A242A acoustic coupler.

I am also grateful to Dribble and Nibble, a.k.a. Slugger Paws and Critter Cat, who offered quiet companionship during many long nights of work.

For Evy ... you know why.

Using this Guide: How and Why

This *Guide* can help you learn the material in Edward Minium and Robert Clarke's textbook *Elements of Statistical Reasoning*. To get the most from the text and this *Guide*, do the following when you set out to study a chapter in the text. (But you need not do this all at one sitting; in fact, you probably shouldn't.)

1. First read the **Objectives** listed in the corresponding chapter of this *Guide*. They'll give you a preview of what's in the chapter and some specific goals to work toward.

2. Also before you begin studying the text, you *may* find it helpful to inspect the **Map** in the corresponding chapter of this *Guide*. The map is an illustration showing how some of the concepts in the chapter interrelate. It too will give you a preview of the chapter—but it may provide too much detail at this time. If it looks complicated, don't let it scare you. When you've worked through the chapter, you'll understand the map.

3. Study the **Chapter** in the text. And as you do so, keep referring back to the objectives. Try to see how each section of the chapter furthers your progress toward one or more of the objectives. In fact, you may find it helpful to construct a kind of index by adding next to each objective the number of any relevant page from the text.

4. Return to this *Guide* and reread the **Objectives**, to fix your goals in your mind.

5. Then do the **Review**. The review will lead you through the important points of the chapter, giving you the opportunity to practice recalling them. Such practice in retrieving information from memory is an important technique for learning the information. You'll also be constructing a summary of the chapter that's more detailed than the one in the text.

In working through a review, fill in the blanks by writing words in the spaces provided in the margin. Consult the list of answers in the back of this Guide, *and make any corrections in the margin. Later on, if you wish to re-review, just cover the margin. This time, jot your answers on a piece of paper, and only after committing them to writing, uncover the margin to check yourself.*

Warning: you will not be able to "cram" successfully by filling in the blanks in a review. Doing a review cannot substitute for careful attention to the text.

6. Study the **Map** for the chapter. As noted above, this is an illustration showing how important concepts in the chapter interrelate. It's essentially a series of sentences; the end of a sentence is shown by an arrowhead. As you inspect a map, then, pay attention to the arrowheads; they tell you where the sentences end.

You may be able to improve on the map, adding information or rearranging what's there. If you can, do so. Working actively with ideas is a fine way to learn them.

Beware: because of limitations on space, the map does not cover all the important points in the chapter. It will help you learn whatever it shows, but there will be more that you should learn from the chapter. Again, you may wish to create your own map of the omitted information.

7. Study any **additional material** in the *Guide* for the chapter you're working on. There are such things as notes on memorization and tricks for catching your errors before your instructor does.

8. Back in the text, work at least some of the **Exercises** the authors have helpfully provided. By giving you practice in using concepts and applying techniques, they solidify your knowledge of the chapter.

9. Return to this *Guide* for the **Self-Test**. *Mark your choices only in the margin, and after you check the answers in the back of the* Guide, *make any corrections only in the margin. In this way, you can test yourself again — as you might wish to do when you review for an examination. Just cover the margin.*

If you do poorly on the self-test, you know you need to work at the chapter some more. But be careful: with its 10 items, the self-test cannot cover all the important points in the chapter. So doing well on a self-test means only that you know *some* of the important points in the chapter—not necessarily all of them. (To identify the important points, use the Objectives and the Review.)

Also available in this *Guide* are one or two sheets of **Homework Exercises** for each chapter. Your instructor may assign these. If not, you can use them as exercises supplementing those in the text. Ask your instructor for the answers.

You may find it helpful to keep track of your progress by making notations on the lists of objectives. To indicate how far you've come in meeting an objective, you could write in next to it something like "Solid," "Okay," "Unsure," or "Ask about this."

CHAPTER 1

INTRODUCTION

OBJECTIVES

Ch. 1 describes the purposes of statistical techniques and introduces some basic concepts and principles. In studying the chapter, work to understand these matters:

Descriptive Statistics
Inferential Statistics
Relationship as a Problem of Description and Inference
Prediction as a Problem of Description and Inference
Population
Sample
  Random Sample
    Random Sampling Variation
Variable
  Quantitative Variable
    Numerically Scaled Variable
  Qualitative Variable
Substantive Question   •   Substantive Conclusion
Statistical Question   •   Statistical Conclusion
Criticisms of Statistics   •   Rebuttals to Criticisms
Tips on Studying Statistics
Exact Number   •   Approximate Number

REVIEW*

What Is (Are?) Statistics?

In ordinary speech, the term *statistics* refers to statements involving numbers, as in the expression "unemployment statistics." The word is plural in this sense ("The statistics *are* due to be released soon"). In Ch. 1, however, the text uses the term in a different sense, to refer to a specialization within the discipline of mathematics. Statistics, in this

*For advice on how to do a Review, see the section on "Using this Guide" at the beginning of this book.

1

sense, is like arithmetic, geometry, and algebra—a set of mathematical techniques for solving problems. In this sense, the word is singular ("Statistics *is* a field of mathematics").

Descriptive and Inferential Statistics

The techniques of the field called statistics are helpful for solving problems that arise in research when an investigator collects observations. There are two kinds of statistical techniques. One kind is called descriptive statistics, and the other kind is called __1__ statistics.

1.

The purpose of __2__ statistics is to organize and summarize observations (data). Calculating an average and drawing a graph are examples of the techniques called __3__ statistics.

2.

3.

The techniques of inferential statistics are helpful when a researcher goes to make inferences from the observations. Chance factors usually play a role in any investigation, and the purpose of __4__ statistics is to take account of the __5__ factors when drawing conclusions. Inferential statistics is also called __6__ statistics.

4.

5.

6.

Relationship and Prediction

Two important problems that frequently arise in research have to do with relationship and prediction. A researcher may wish to know how one characteristic of his or her subjects is related to another characteristic of the subjects. For example, the researcher may wish to know how strongly a person's score on a certain test is related to that person's score on the same test when it is readministered four months later. If the score is influenced by a stable characteristic like the person's intelligence, there will be a strong relationship between the first score and the second. That is, people who scored high the first time will generally score high the second time, while people who scored low the first time will generally score low the second. But if the score is markedly influenced by an unstable characteristic such as the subject's current mood, there will be little __7__ between the first score and the second. To assess the extent of the relationship between one characteristic and another, a researcher will collect data and apply __8__ statistics to organize and summarize the data. To reach a conclusion, the researcher will apply __9__ statistics.

7.

8.

9.

Questions about prediction arise when a researcher knows one characteristic of the subjects and wants to use that knowledge to estimate another characteristic. For example, a personnel office may administer an aptitude test to people seeking clerical work, and the office may hire only people who earn a sufficiently high score on the test. But does the score really allow an accurate prediction of how well the person will do on the job? To answer such a question, a researcher must collect observations, use descriptive statistics to 10 and 11 the observations, and then apply inferential statistics to draw a(n) 12 from the observations.

Problems involving relationship and prediction are thus matters to which an investigator can still apply the two kinds of statistical techniques, namely 13 and 14 statistics.

10.

11.

12.

13.

14.

Populations, Sampling, and Samples

A population is the complete set of observations about which a researcher wishes to draw conclusions, and a sample is a 15 of a population. Techniques of description can be applied either to a 16 or to a sample, but techniques of inference are applied only to a sample, in order to derive from that sample a conclusion about the population from which it came. The inferential techniques take into account the 17 factors that operated when the sample was selected from the population. If we could inspect a complete population, there would be no need to apply inferential techniques, because we could learn everything we would want to know about the population by simply applying the other techniques, those of 18 .

15.

16.

17.

18.

The process of selecting a sample from a population is called sampling. There are many possible ways to select a sample, but in statistical work, one particular kind of sampling is especially important. This is random sampling, and it is sampling in which 19 does the selecting of the observations. The chance factors that produce a given 20 sample will vary from sample to sample, of course, and thus the samples themselves will vary. This chance variation from sample to sample is known as 21 sampling 22 .

19.

20.

21.

22.

Suppose there is a population of observations from which we draw a random sample, and suppose the sample is small in size (consisting of just ten observations, say). If we draw additional samples of this same small size from the given population, the random 23 variation across these samples will probably be relatively large.   That is, there will probably be major differences among the samples.

23.

But now suppose that from the same population we draw a random sample that is large in size (consisting of, say, 100 observations). If we select additional samples of this same large size from the given popula-

tion, the random sampling __24__ across the samples will probably be
relatively __25: large or small?__ . That is, there will probably be only minor
differences among the samples.

24.
25.

In general, then, when the size of several random samples from a given
population is larger, the random sampling variation among those samples
is __26: larger or smaller?__ .

26.

Variables

In the course of an investigation, a researcher will make observations on
subjects, focusing on one or more of their characteristics (for example,
their political affiliation—Democrat, Republican, or Independent—or
their age). A characteristic that may vary from subject to subject, taking
on different values, is called a __27__ .

27.

There are different kinds of variables. A __28__ variable is one whose
values differ in quality, or kind, rather than in amount. Political affiliation is an example of a __29__ variable, in that the values of this variable—
Democrat, Republican, or Independent—differ in kind. A Democrat is
not more or less of something than a Republican; a Democrat is a person
who differs in quality from a Republican. In contrast, a characteristic
whose values differ in quantity or amount is called a __30__ variable. Age is
an example of a __31__ variable.

28.

29.

30.
31.

There are some quantitative variables whose values are numbers that
represent the quantities of the characteristic that varies. Such variables
are said to be __32__ scaled. Age is an example.

32.

Other quantitative variables have values that differ in quantity but are
not represented precisely by numbers—for example, rank in a military
hierarchy. The rank of sargeant is more of something (it's more distant
from the bottom of the hierarchy) than the rank of corporal, but there are
no numbers here to tell us the exact degree of the varying characteristic.

The textbook is concerned mostly with quantitative variables that are
numerically __33__ , but it also offers information on the treatment of
variables whose values differ in kind (__34__ variables).

33.
34.

How Does Statistics Figure Into Research?

Any investigation begins with a question about the subject matter of the
investigation. Such a query is called a __35__ question. During the investigation, the researcher will collect observations, and she or he will use the
techniques of __36__ statistics to organize and summarize the observations
(the data). The researcher may calculate an average of the scores that she

35.

36.

or he collected, for example. The researcher now asks a question about this average or whatever statistical index is in use. For example, the researcher may ask, "What is the likely value of the average score in the whole population from which I sampled?" The answer to such a query is a __37__ conclusion. But such a conclusion is not yet the end of the study. The researcher must still do some careful reasoning to determine the larger meaning of the statistical __38__. On the basis of the statistical conclusion, but also on the basis of other considerations, the researcher will reason out a conclusion about the subject matter. Such a conclusion is called a __39__ one.

37.

38.

39.

Exact Numbers and Approximate Numbers

Some numbers are exact. A careful count of the number of subjects in an experiment, for example, is a(n) __40__ number, since it contains __41:__ __some or no?__ margin of error. But most numbers generated by a measuring procedure, such as heights and weights, lack this kind of accuracy. They are known as __42__ numbers.

40.

41.

42.

An approximate number can be calculated to many decimal places, and the question arises of how many decimal places to figure and to report. No simple answer to this question is possible. You should not calculate a(n) __43__ number to many decimal places more than the number in the figures that entered into the calculation. But if you do excessive rounding, you will lose the accuracy to which you are entitled. The examples in the text will provide good models. Follow them.

43.

ARE YOU WORRIED ABOUT THE MATHEMATICS IN THIS COURSE?

As of this writing, I have led almost 700 students through an introductory course in statistics. Nearly all of them, I'm sure, were initially worried about the mathematics facing them. One young woman even dreamed of being attacked by numbers on the night before her first class with me. But 95% of these students passed the course with a grade of *C* or better, in spite of my high standards, and the person who had the nightmare earned a strong *A*. She also acquired good self-confidence. *You can be a success too.*

To put your mind at rest about the mathematics in this course, consider these facts:

- Your text emphasizes the *logic* of statistics, not the theorems, formulas, and proofs that mathematicians work with. The title of the text is *Elements of Statistical **Reasoning**,* and it's reasoning, not mathematics, that's important here.

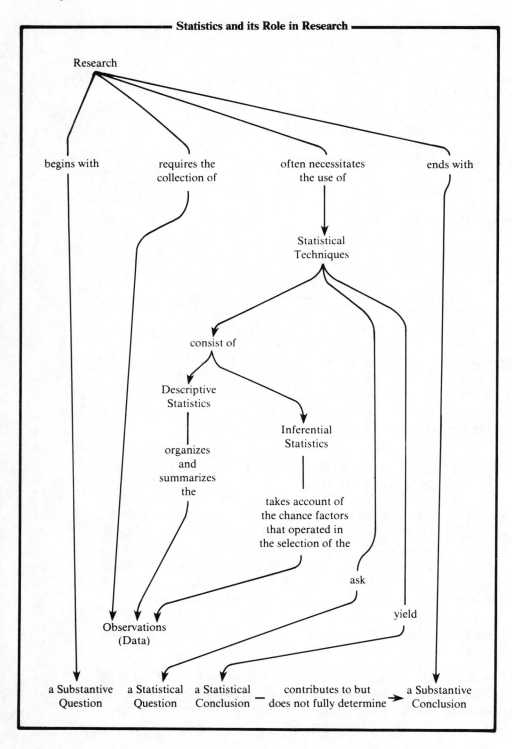

- The mathematics employed in the text is only simple algebra. You covered this in high school, and if you need to relearn it, you can do so easily. Appendix A in the text will help.

Furthermore, look what you've got going for you:

- Your text presents much more than the bare facts. It also provides the "big picture," so you can see how the facts fit together. It also adds the details, so you can gain insight into the facts.
- Right under your nose at this very moment is a study guide that offers you a lot of help with the text.
- Your instructor (and your teaching assistant, if you have one) will go over the material in the text and will answer any questions you have.

Moreover:

- *This course itself provides a leisurely review of mathematics.* The math in the course begins with counting (tallying up observations). It goes on to proportions and percentages, and more complex matters come up only later. So you can gradually relearn whatever you're uncertain of—and you'll be relearning it in a context that makes it vivid and useful.

The mathematics in this course is thus fully within your comprehension. If you've got adequate time to devote to the course, you can learn absolutely everything in your text and feel really good about it.

TIPS ON BUYING A CALCULATOR

A miniature calculator would be a good investment for this course, and you'll probably find other uses for it too.

You don't need anything fancy. You'll have no use for the special features of the "scientific" calculators that are meant to replace a slide rule—no use for the keys for pi (π), logarithms, exponentiation, or the trigonometric functions sine, cosine, tangent, and cotangent. (As noted above, this course does not require sophisticated mathematics.) You do want the following features:

- An add-on memory, which permits you to add a number to another number already in storage. The key for doing this is usually labeled $M+$. A machine with add-on memory typically has other keys labeled $M-$ (for subtracting something from what's in memory), MR or RM (for recalling what's in memory), and MC or CM (for clearing what's in memory). These keys will all be useful.
- Automatic constant for multiplication (for which there's no special key), or a key for squaring a number, labeled x^2. Automatic constant works like this: to square

a number (to multiply it by itself), you first enter the number and then press the $\times$ key. Instead of entering the number again, though, you just press the = key. See if this works on any calculator you're trying: $2\times$ = should get the calculator to read 4 (2×2); $3\times$ = should get the calculator to read 9 (3×3); and so on. A key labeled x^2 is even better for this same purpose.

You should also look for a machine with:

- Positive-action keys, which click or change in amount of resistance to the touch when they work. Some machines give you no feedback on whether a key has functioned when you depress it, and you then must constantly check the display, which is a nuisance.
- Keys that are big enough for you to press easily and accurately. Too much miniaturization is a liability.
- A square-root key. It'll be labeled $\sqrt{}$ or $\sqrt{x}$.
- A display that you can read easily from a variety of angles.

Be a smart shopper:

- Compare models, guarantees, and prices, and try several stores.
- Ask about the stores' policies on defective merchandise. What will they do if your purchase malfunctions after you get it home? The store should agree that if it proves defective within 30 days, they will replace it with a new machine from their own stock, rather than send the old one to the factory for repair. Ask the salesperson to write "30-day exchange" on your receipt and sign it.

You should be able to get what you want for less than $20.

If you're willing to spend a little more money, look for a calculator with features expressly intended for statistical use. In advertising, these are often called statistical functions. Such a calculator usually has keys labeled Σ (a Greek symbol for summation) or $\Sigma+$, $\overline{X}$ ("eks bar," a symbol for a type of average) or MEAN, and σ ("sigma," a symbol for an index of variability). These keys can save you some drudgery.

Even fancier—and more expensive—calculators offer features useful in studying relationship and prediction. The relevant keys are usually labeled r (a symbol for an index of correlation) or CORR (for *correlation*) or L.R. (for *linear regression*—you'll understand it when you get to Ch. 8). These features too can spare you some tedious work.

If you buy a calculator, take the time soon to learn the basics of it: how to add, subtract, multiply, and divide, and how to use the memory. If the machine has additional features that now seem puzzling, don't worry. You can learn to use them later, when they'll be helpful in this course.

HOW TO DO YOUR BEST ON A
MULTIPLE-CHOICE TEST

1. Underline the important words in the questions and again in the possible answers. Be sure to underline negatives such as *none* and *not*.

2. Mark each choice for a given question *T* (for *true*), *F* (for *false*), or *?* (for *unsure*).

3. If you find an item especially difficult, leave it for later.

4. Answer all items. Guess if you have to.

5. If you must guess, eliminate answers you are sure are wrong before you do so.

6. If you have second thoughts about your answer for a given item and are considering changing the answer, do change it if you think you ought to.

The last point deserves comment. A persistent myth among students says that you're better off sticking with your first choice on a multiple-choice item, but educational psychologists have demonstrated that a majority of students generally improve their scores when they change their answers.*

Now, are you one of the majority who're better off changing, or one of the minority who should stick with the first choice? Right now, you don't know—but it's clear what you should do: change an answer whenever you think you ought to. The odds are in your favor if you change answers, because you're more likely to be one of the majority than one of the minority. This is an illustration of the point on p. 8 of your text: "The 'group' approach can often be turned to the probable advantage of an individual."

But keep track of what happens each time you make a change: did you change a wrong answer to a right one, a right to a wrong, or a wrong to another wrong? After you have a good-sized sample of cases in which you changed the answer—about 25 such cases—you can tell with some confidence whether you're really in the majority, or whether you're among the minority of students who generally lower their scores when they change answers.

And illustrated here is another principle from Ch. 1 of the text: "Large samples will provide a more precise estimate of what is true about the population than can be expected from small samples" (p. 5). The population in this instance is the complete set of answer changes you might ever make, and every time you change an answer, you add another case to the sample you're collecting. In accord with the principle just quoted from the book, don't try to generalize from a small sample of answer

*See, for example, "Answer-Changing Behavior and Grades" by J. J. Johnston in *Teaching of Psychology* for February 1978 (Vol. 5, No. 1).

changes to the complete population; a small sample is too likely to be unrepresentative of the whole population.

SELF-TEST

For each question, choose the one best answer. Check your choices by consulting the list of correct answers at the back of this book. As noted on the page on "Using this *Guide*," you should mark your choices and correct them only in the margin. Then to retest yourself, you need only cover the margin.

The type of statistics that organizes and summarizes data is: A) sampling statistics. B) descriptive statistics. C) inferential statistics. D) the statistics of approximate numbers.

1.   A B C D

Another name for inferential statistics is: A) descriptive statistics. B) sampling statistics. C) exact numbers. D) approximate numbers.

2.   A B C D

The purpose of inferential statistics is to take into account: A) the lack of accuracy in approximate numbers. B) the lack of numerical scaling in qualitative variables. C) the chance factors associated with sample selection. D) the depersonalization of descriptive statistics.

3.   A B C D

Calculating the proportion of subjects who are female is an example of: A) drawing a substantive conclusion. B) sampling statistics. C) inferential statistics. D) descriptive statistics.

4.   A B C D

In statistical work, the term *population* is defined to refer to: A) people. B) subjects. C) numbers. D) observations.

5.   A B C D

A population is a(n) _____ set, and a sample is a(n) _____ set. A) exact ... approximate B) complete ... partial C) approximate ... exact D) partial ... complete

6.   A B C D

A sample selected by a process of chance is a(n) _____ sample. A) random B) large C) exact D) numerically scaled

7.   A B C D

In comparison to random sampling variation among small samples, random sampling variation among large samples is: A) greater. B) unchanged. C) smaller. D) less exact and more approximate.

8.   A B C D

An example of a numerically scaled variable is: A) sex. B) political affiliation. C) hair color. D) the count of the errors on a test such as this.

9.   A B C D

The last step in an investigation is the: A) statistical conclusion. B) statistical question. C) substantive conclusion. D) substantive question.

10.   A B C D

CHAPTER 2

FREQUENCY DISTRIBUTIONS

OBJECTIVES

Ch. 2 presents the basic techniques of description for a jumble of numbers; they are tables of certain kinds. In studying the chapter, work to understand these symbols, terms, concepts, principles, and techniques:

Frequency Distribution for Ungrouped Scores
   f • n
The Grouping of Scores
   Class Intervals   •   Score Limits   •   Benefit of Grouping   •   Liabilities of Grouping   •   Principles for Forming Class Intervals
Frequency Distribution for Grouped Data
   Range
Relative Frequency Distribution
   Proportionate Frequency (*prop f*)   •   Percentage Frequency (%*f*)
Exact Limits
   The Cumulative Percentage Frequency Distribution
    cum f   •   *cum* %*f*

Also in studying Ch. 2, work to learn these skills:

Construction of a Frequency Distribution for Ungrouped Scores
Construction of a Frequency Distribution for Grouped Scores
Extension of a Frequency Distribution to Show Relative Frequency
Construction of a Cumulative Percentage Frequency Distribution

REVIEW

The Frequency Distribution

Observations as a researcher collects them typically accumulate in a jumble. A first listing of the observations thus shows only an unorganized collection of scores. (Look at Table 2.1 on p. 14 of your text for an example.) The purpose of descriptive statistics is to organize and summarize a jumble like this, and the most fundamental way to accomplish this purpose is to construct a frequency 1 , which is a kind of 2: graph or table? .

1.

2.

Whether a collection of observations is a sample or a population, the technique of constructing a frequency distribution for it is the same.

To construct a frequency distribution (with no grouping of the scores), locate the highest and the 3 score values that actually occurred. Then record *all possible* score values, including the highest and the lowest that occurred and all values in between (even if some of the in-between values did not show up in the data). Record the possible score values in 4: ascending or descending? order down the page. Then add a second column to the right of the first one. In this column, list for each possible score value its frequency, which is the number of times it occurred. (Look at Table 2.2 on p. 15 for an example. In this table, as in Tables 2.3 and 2.4, *frequency* is written out in full. Thereafter, *frequency* is abbreviated 5 , as in Table 2.5.)

3.

4.

5.

The term *distribution* is sometimes used to refer to a collection of scores, and the term is appropriate, as you can see when you inspect a collection of scores that has been recast into a table of the kind you are studying. The scores that occurred are a distribution in the sense that they are distributed over the range of possible values, and the table shows exactly how.

The Grouped Frequency Distribution

When the scores cover a wide range (as in the data shown in Tables 2.1 and 2.2), the left-hand column of the frequency distribution is very long. In such a case, combining individual score values in the left-hand column to form a smaller number of groups makes it easier to display the data and to grasp their meaning. The right-hand column must then list the frequency with which scores occurred within each given grouping. A table constructed in this way is called a 6 frequency distribution. (Look at Table 2.4 on p. 16 for two examples of how the data in Tables 2.1 and 2.2 can be grouped and displayed in a 7 distribution.)

6.

7.

Disadvantages of Grouping Scores

Though it is often helpful to group scores in constructing a frequency distribution, this technique has two disadvantages. First, a person who sees only a 8 frequency distribution does not get all the information

8.

that is available in the original collection of scores. Thus the person cannot know *exactly* what values the original scores took. Second, a collection of individual scores does not yield a unique collection of 9   scores. That is, there is more than one way to construct a 10 distribution. Thus a researcher, knowingly or unknowingly, may select a system of grouping that is misleading, giving a false impression of the original collection of scores.

9.

10.

Guidelines for Constructing a Grouped Frequency Distribution

A grouping of possible score values is called a class 11 . The highest and the lowest of the possible values in a 12 interval are called the score 13 . The number of values within a grouping is called the width of the interval. Thus in the grouped frequency distribution shown on the left side of Table 2.4 on p. 16, the intervals have a width of three, because each interval consists of three values. (For example, the interval 96 - 98 covers the three scores 96, 97, and 98.)

11.

12.

13.

In determining how to set up a system of class intervals for a given collection of scores, a researcher usually follows seven rules:

1. The class intervals 14: should or should not? overlap one another.

14.

2. The intervals should be of 15: the same or varying? width.

15.

3. The intervals should cover all possible scores from the highest that actually occurred to the lowest. That is, the intervals should be 16 throughout the distribution, leaving no gaps.

16.

4. The interval containing the highest score should be placed at the 17: top or bottom? of the table.

17.

5. There should be between 10 and 18 class intervals.

18.

6. The width of the intervals should be a value such as 2, 5, 10, or 25, a value that is 19 to work with in doing arithmetic.

19.

7. The lower score limit of each interval should be a 20 of the interval width. For example, if the width is 5, then the lower score limits should be the numbers that result from multiplying 5 by a series of digits, numbers such as 30, 35, 40, 45, and so on.

20.

The Relative Frequency Distribution

Scores as they are originally recorded are sometimes called raw scores, and counts of raw scores are called raw frequencies. It is often helpful to "cook" raw frequencies by transforming them into relative frequencies. A relative frequency is a frequency that is relative to the total number of scores in the collection. A table that shows relative frequencies for a given collection is called a __21__ frequency distribution. (Look at Table 2.7 on p. 21 of the text for an example.)

There are two kinds of relative frequencies: proportions and __22__ (both are illustrated in Table 2.7). A proportion is symbolized *prop f* in Table 2.7. To calculate the proportion of cases in a given class interval, divide the frequency count for that interval by the total __23__ of cases in the distribution. In symbols, this instruction is to calculate __24__, where __25__ is the symbol for a frequency count, and __26__ is the symbol for the total number of cases.

A percentage frequency is symbolized *%f* in Table 2.7. To calculate the percentage of cases in a given class interval, first find the proportion and then multiply it by 100. Multiplying a number by 100 is easy to do. Rather than writing out the computation (or punching it into a calculator), just move the decimal point in the number to be multiplied by 100 two places to the __27: left or right?__ .

If you want to compare two or more frequency distributions in which the number of cases differs from one distribution to another, looking at the raw frequencies is not very helpful. It is better to compare the __28__ frequency distributions.

21.

22.

23.
24.
25.
26.

27.

28.

Exact Limits

Exactly what does a number like a score of 87 on a test mean? Instead of presuming that it indicates precisely 87.00000... units of some quantity, it is more reasonable to take the score as representing a quantity that is closer to 87 than to 86 or to 88. Thus we can best treat a score of 87 as representing a range of values from 87.5 to 88.5 The values 87.5 and 88.5 are called the exact __29__ of the score of 88.

In this example, we have assumed that scores on the test were measured to the nearest "point," or whole number. The __30__ limits of the score of 87 thus extend from one-half of a point below the score to one-half of a point above it. In general, the exact __31__ of a score are considered to extend from one-half of the smallest unit of __32__ below the value of the score to one-half of this unit __33__ the value of the score.

In a grouped frequency distribution, it is helpful to indicate the class intervals in terms of the exact limits of the scores involved. (The second column of Table 2.8 on p. 24 provides an example.) The lower exact limit of a given class interval serves at the same time as the upper exact limit of the class interval immediately __34: above or below?__ it.

29.

30.

31.
32.
33.

34.

The Cumulative Percentage Frequency Distribution

It is often useful to know the percentage of cases falling below a particular point along the range of possible scores. A table showing the percentage of cases lying below the upper exact limit of each class interval is called a 35 percentage frequency distribution. (Look at Table 2.8 for an example.)

To construct a cumulative percentage frequency distribution, we must start with an *un*cumulated frequency distribution—which is simply the kind that we have been considering up to now. In an additional column, we begin at the bottom and record the total frequency falling below the 36: upper or lower? exact limit of each class interval. Such a number is called a cumulative frequency and is symbolized 37 in Table 2.8. If we do our arithmetic correctly, the cumulative frequency for the uppermost class interval will equal the total number of cases, which is symbolized 38 . Finally, each cumulative frequency is converted to a cumulative percentage frequency. To make this calculation, divide the cumulative frequency, symbolized 39 , by the total number of cases, symbolized 40 , and multiply the result by 100 (which is equivalent to moving the decimal point two places to the 41: left or right?).

35.

36.
37.

38.

39.
40.
41.

The Frequency Distribution for Qualitative Data

A frequency distribution can be constructed not only for scores (which are 42: qualitative or quantitative? data of the 43 scaled kind), but also for *qual*itative data. In the case of qualitative data, each observation is a categorization. It indicates that a certain subject falls into a certain category. (For example, in a study of the majors that students choose, we may observe that the subject named Lena Learner falls into the category Chemistry Major.) A frequency distribution for observations of this kind consists of two columns. (Look at Table 2.3 on p. 15 for an example.) The left-hand column lists the various possible categories, and the right-hand column lists the frequency count for each category—that is, the 44 of subjects who were classified as falling into the given category.

42.
43.

44.

Because the various categories do not indicate different degrees of some quantity, there is no convention governing the listing of the categories in the left-hand column. (In contrast, for quantitative data, it is conventional to list the scores or class intervals in numerical order with the largest value at the 45: top or bottom? .) A common choice is to order the categories by their frequencies, with the category whose frequency count is largest at the top. (This was done in Table 2.3. The category

45.

The Frequency and Cumulative Frequency Distributions

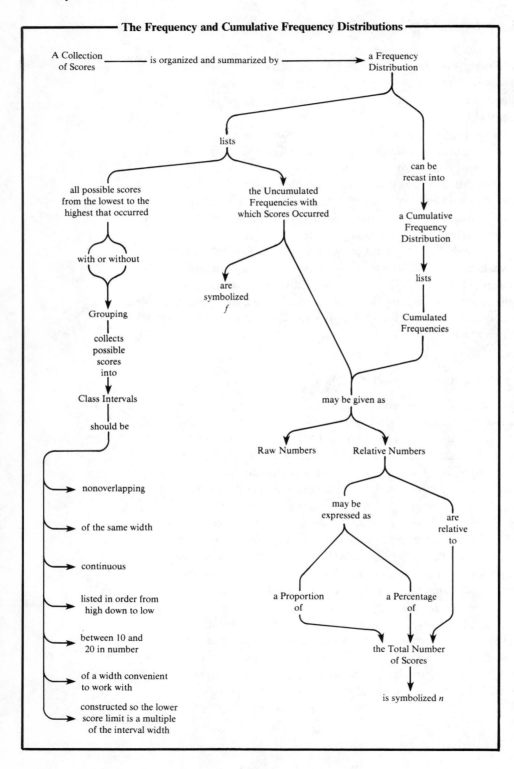

Biology was put at the top because its frequency is the highest; Geology was put at the bottom because its frequency is the lowest.)

The raw frequencies for a distribution of qualitative data can be converted to relative frequencies, as for a distribution of scores, but it is not logically possible to construct a cumulative frequency distribution for qualitative data.

EXERCISES

Here are some exercises to supplement those in your text. The ones in the text are well worth doing, but these are different and will help you gain insight into the kinds of tables that Ch. 2 presents.

Consider the table immediately below. Never mind what the scores might mean; that's irrelevant for now. Just think about the internal logic of the table. Only the top value in the column of cumulative frequencies and four figures in the column of cumulative percentage frequencies appear here, but from these five numbers you can determine all the missing information. Just recall what you know about the topmost value in the *cum f* column, and remember how you go about finding the *cum %f* figures.

Give this problem a good, honest try. You'll feel really proud if you solve it for yourself.

Score Limits	Exact Limits	f	cum f	cum %f
18–19	17.5–19.5		12	
16–17	15.5–17.5			75
14–15	13.5–15.5			50
12–13	11.5–13.5			33
10–11	9.5–11.5			25

If you think you can't figure out how to get the numbers that are missing in the table, look at the hint below* and then try again.

A complete version of the table appears at the back of this book. Use it to check your work.

---

*The topmost *cum f* value always equals *n*. If *n* is 12, then *cum f* for the bottommost class interval is 25% of 12, or 3. And 3 must also be the *f* value for the bottommost class interval. Fill in these numbers and keep working on getting the rest.

Below are two more problems of the same kind. Working them will help you understand the connections among the various numbers in tables like these.

Score Limits	Exact Limits	*f*	*cum f*	*cum % f*
30–34			12	
25–29				67
20–24				50
15–19				33
10–14				33

Score Limits	Exact Limits	*f*	*cum f*	*cum % f*
120–129			15	
				67
–109		2		
				33

Here are some principles that describe the connections among the various parts of tables like these. Note how the principles are illustrated in the tables above (when the numbers are correctly filled in).

1. The sum of the *f* values is *n*. Why? Each *f* is a count of a part taken from the whole collection of scores. When we sum (put together) the counts of all the parts, we have the count of the whole, which is symbolized *n*.

2. The top figure in the *cum f* column is *n*. By definition, the top figure in this column is the count of the number of scores that fall below the upper exact limit of the class interval containing the largest score—and all *n* scores fall below this upper limit.

3. The top number in the *cum % f* column is always 100. This figure says that 100% of the scores—all of them—fall below the upper exact limit of the top class interval.

4. All figures in the column of *f* values and all figures in the column of *cum f* values must be *whole* numbers without any fractional part (0, 1, 2, 3, and so on), because these figures are counts, counts of the numbers of scores of various kinds—and there is no such thing as a fraction of a score.

5. The figures in the $\%f$ and *cum* $\%f$ columns, as percentages, must be between 0 and 100.

6. A given figure in a cumulative column is the sum of two numbers: the cumulative figure immediately below, and the corresponding uncumulated value in its row to the left.

7. In the bottom row, the figure in a cumulative column is the same as the figure in the corresponding uncumulative column.

8. As you go up a cumulative column, the numbers must either stay the same from one row to the next or increase. The numbers cannot become smaller.

Note that the three tables in these exercises are unrealistic, in that they show too small a number of class intervals. Frequency distributions should list between 10 and 20 class intervals.

SELF-TEST

A frequency distribution is: A) a graph.   B) an unorganized collection of scores.   C) a count of the total number of scores.   D) a table.
    1.   A B C D

The left-hand column of the ungrouped frequency distribution for a collection of scores should list: A) just those scores that occurred, with the highest on top.   B) just those scores that occurred, with the lowest on top.   C) all possible scores between the lowest and the highest that occurred, with the highest on top.   D) all possible scores between the lowest and the highest that occurred, with the lowest on top.
    2.   A B C D

Fundamentally, a frequency is: A) an approximate number.   B) a count.   C) a fraction.   D) the width of a class interval.
    3.   A B C D

In a grouped frequency distribution, the top and bottom scores in a given grouping are: A) the range of the class interval.   B) the exact limits of the class interval.   C) the value of f for the class interval.   D) the score limits of the class interval.
    4.   A B C D

Suppose the lowest score in a sample of observations is 8. If we decide to summarize the sample in a grouped frequency distribution with class intervals that are 3 units wide, the best choice for the bottommost class interval is: A) 6 to 8.   B) 7 to 9.   C) 8 to 10.   D) 9 to 11.
    5.   A B C D

The letter n is the symbol for: A) the number of class intervals.   B) the number of observations.   C) the number of possible scores between the lowest and the highest that actually occurred.   D) the width of a class interval.
    6.   A B C D

In the fall semester, 27 students in a statistics class took the final exam. In the spring semester, the same professor taught the same course with the same final, and 38 students took the exam. To compare the scores from the first semester with the scores from the second semester, the *worst* choice would be to display each collection of scores in a frequency distribution showing: A) raw frequencies.*   B) proportions.   C) percentages.   D) relative frequencies.

7.   A   B   C   D

To express a frequency as a proportion, one should calculate: A) n/f. B) $(n/f) \times 100$.   C) f/n.   D) $(f/n) \times 100$.

8.   A   B   C   D

When measured to the nearest 10 feet, the height of the Sears Tower in Chicago is 1450 ft. The figure 1450 here represents values between: A) 1449.5 and 1450.5.   B) 1449 and 1451.   C) 1445 and 1455.   D) 1440 and 1460.

9.   A   B   C   D

At the left of the middle row of a cumulative frequency distribution is the entry "20 - 24". The cumulative frequency in this row is 88. For the collection of scores summarized in this table, there are 88 scores that fall below the value: A) 24.5.   B) 24.0.   C) 22.5.   D) 20.0.

10.   A   B   C   D

*The term *raw* is explained in the review of this chapter under the heading *The Relative Frequency Distribution.*

CHAPTER 3

GRAPHIC REPRESENTATION

OBJECTIVES

Ch. 3 presents four techniques for describing observations by means of a graph. These techniques are alternative ways of displaying the information given in tables of the type introduced in Ch. 2. In studying Ch. 3, work to understand these symbols, terms, concepts, principles, and techniques:

Principles of Constructing a Graph
    X Axis   • Y Axis
Histogram
Bar Diagram
Frequency Polygon
    Midpoint of Class Interval
Cumulative Percentage Curve (Ogive)
    Centile   •   Centile Rank   •
    Percentile   •   Percentile Rank   • C_{50}
Proportion of Area Equals Relative Frequency of Cases
The Effect of Sampling Variation on the Shape of a Frequency Distribution
There is No Such Thing as *the* Graph of a Set of Data
Characteristics of a Frequency Distribution
    Central Tendency
    Variability
    Shape
        J-Shaped • Skewed • Rectangular • Bimodal • Normal

Also in studying Ch. 3, work to learn these skills:

Construction of a Histogram
Construction of a Bar Diagram
Construction of a Frequency Polygon
Construction of a Cumulative Percentage Frequency Curve
Determination of a Centile from a Graph
Determination of a Centile Rank from a Graph

REVIEW

The Histogram

Look at Table 3.1 on p. 30 of your text. The table presents the
1: grouped or ungrouped? frequency distribution for a collection of
2: how many? scores. Now look at Figure 3.1 on p. 31. This is a graph of
the same information, and it is called a 3 . The graph consists of a series
of rectangles, each of which represents the scores that fall within a certain
class interval. A given rectangle is erected so that its two vertical
boundaries coincide with the 4: score or exact? limits of the interval. The
height of the rectangle represents the frequency with which scores occurred in that interval. The frequency can be either the raw number (the
simple count of the scores in the interval) or the relative frequency (the
proportion or the 5: what is the other kind of relative frequency?) of scores
in that interval.

A histogram should be constructed in accord with six principles:

    1. It is customary to represent scores along the 6: horizontal
or vertical? axis (the 7: X or Y? axis) and to represent raw or relative
frequency along the 8: horizontal or vertical? axis (the 9: X or Y? axis).

    2. The intersection of the two axes should be at the 10 point on both
scales. If it is not, the reader should be warned with a device like the
break in the horizontal axis of Figure 3.1.

    3. If values are to be read from a graph, the larger the scale, the better.

    4. To identify positions along the axes, use convenient numbers. On
the horizontal axis, the limits of the class intervals or the midpoints of the
intervals usually 11: do or do not? form the most convenient frame of
reference.

    5. The height of the histogram should be about 12: what fraction? of
the width.

    6. The graph should have a succinct, informative title, and both axes
should have 13 .

1.

2.

3.

4.

5.

6.

7.

8.

9.

10.

11.

12.

13.

To represent relative frequency instead of raw frequency in a histogram (or any other graph), it is not necessary to redraw the rectangles (or
whatever makes up the figure). It is necessary only to relabel the vertical
axis. Figure 3.1 as it appears in the text, for example, shows raw
frequencies. To display relative frequencies—proportions, say—the point
labeled "5" should be relabeled with the proportion corresponding to a

frequency count of 5. Since there were 100 scores in the distribution, this is simply 5/100, or .05. The point labeled "10" should be relabeled with the proportion corresponding to 10, which is 10/100, or .10. And so on.

The Frequency Polygon

The histogram is one way to graph a frequency distribution, and the frequency __14__ is another. In this kind of graph, we make use of the __15__ of the class intervals. To find the midpoint of an interval, locate the point __16__ way between the two score limits (or, alternatively, half way between the two __17__ limits). Over the midpoint of a given class interval, we plot a point at a height that represents the __18__ of the scores in that interval. Again, this can be either the raw or the __19__ frequency. Then we simply connect the dots with straight lines.

14.
15.
16.
17.
18.
19.

If we do nothing further, we will have a graph consisting of a zig-zagging line that does not touch the horizontal axis at either end. It is customary to bring the line down to the horizontal axis, though, and to do this, we identify the class interval just below the one in which the lowest score occurred. The frequency count (raw or relative) for this interval is, of course, zero. We plot the point for this class interval over its midpoint at a height of __20__ , and we connect this point to the point for the next class interval up. Then we do the analogous thing at the upper end of the distribution. In effect, we show in the graph that zero scores occurred in the class interval represented at the left of the __21__ polygon, and that __22__ scores occurred in the class interval represented at the right of the frequency __23__ . Figure 3.3 on p. 33 provides an example.

20.

21.
22.
23.

The Cumulative Percentage Curve

As you learned in the previous chapter, a frequency distribution (a kind of __24: graph or table?__) can be extended to show the cumulative frequency distribution, which is also a __25: graph or table?__ . An example appears in Table 3.2. The cumulative frequency __26__ can be graphed as a cumulative percentage __27__ , such as Figure 3.4 on p. 34.

24.
25.
26.
27.

In the cumulative __28__ curve, the horizontal axis again shows the various values for the scores in the distribution. The vertical axis again shows frequencies, but here these are a particular kind of relative frequency, namely __29: proportions or percentages?__ , and these are cumulative values that start at zero and run to the maximum figure, which is __30__ %.

28.

29.

30.

A cumulative percentage frequency is the percentage of scores that lie below the _31: upper or lower?_ exact _32_ of a given class interval. Therefore, in graphing such a number, it is necessary to plot it over the location on the horizontal axis that represents the _33: upper or lower?_ exact limit. The points are connected with straight lines.

Another name for the cumulative percentage curve is the _34_ . Strictly speaking, this term implies that the curve (which is really a zig-zagging line) has the shape of an **S**. The graph of a cumulative frequency distribution will tend toward this shape when there are more cases in the _35: center or ends?_ of the *un*cumulated distribution than elsewhere.

31.
32.
33.

34.

35.

Centiles and Centile Ranks

The centile system is widely used in educational and psychological measurement to report the standing of an individual relative to the performance of a known group. If we have the distribution of scores for the entire group, we can use the centile system to say where the score for a given individual falls within this distribution. We simply report the centile rank for the individual's score. The centile rank (also called the _36_ rank) for the given score is the percentage of cases that fall _37: above or below?_ the score. In Table 3.1 on p. 30, for example, you can see that eight scores lie below the value 69.5. These eight scores constitute 8% of the whole distribution. Thus in this particular distribution, the centile rank of a score of 69.5 is 8.

36.
37.

Another way to report the same information is to start with the cumulative percentage. Name the cumulative percentage and then say what score cuts off this percentage of the distribution when you count up from the bottom. In the example just cited, the bottom 8% of the scores fall below the value 69.5. In terms of the centile system, the eighth centile is 69.5.

In general, a centile is a point along the scale of scores. The centile (also called the _38_) is identified in terms of the percentage of scores that lie below it. The tenth centile is whatever point cuts off the bottom 10%; the twentieth centile is whatever point cuts off the bottom _39_ %; and so on.

38.

39.

In symbolic form, $C_8 = 69.5$. The symbol C stands for a _40: centile or centile rank?_ . The subscript names the corresponding _41: centile or centile rank?_ . (The symbol C_8 is read "see-sub-eight.")

The _42: which graph?_ can be used to find the centile rank of a score or the centile corresponding to a given centile rank.

40.
41.

42.

Comparison of Different Distributions

Two or more collections of scores can be compared by means of their uncumulative or cumulative frequency distributions, as you learned in the previous chapter. They can also be compared by means of their graphs (the histograms, the frequency __43__ , or the cumulative __44__ curves). When the distributions have unequal numbers of cases, comparison is easiest if the graphs show __45: raw or relative?__ frequencies.

43.

44.

45.

Proportion of Area Equals Relative Frequency of Cases

In a histogram, a rectangle represents a certain number of cases from a particular distribution. The entire set of rectangles represents the entire collection of cases in that distribution. Any given rectangle, as a geometric figure, constitutes a certain proportion of the entire area of the histogram (a certain proportion of the entire area of the set of rectangles). And the cases represented by the rectangle constitute a certain proportion of the entire number of cases in the distribution.

There is a simple relationship between the proportion of area taken up by a rectangle and the proportion of cases represented by the rectangle. The two proportions are __46__ . In general, the proportion of the area of the histogram that lies between one score point and another equals the relative __47__ of the cases with values between those points. This equality is also true for a __48__ polygon.

46.

47.

48.

Sampling Variation in Shape

If we collect a very large number of scores, and if we organize and summarize them in a frequency distribution and then graph that frequency distribution as a histogram or as a frequency polygon, the shape of the histogram or the frequency polygon will probably be markedly __49: regular or irregular?__ . But when we sample just a few cases from such a collection, the shape of the resulting distribution will probably be __50: regular or irregular?__ . The fewer the cases in the sample, the __51: greater or smaller?__ the irregularity is likely to be.

If we want our graph to show whatever regularity exists, we can use a technique that tends to suppress irregularity. One option is to use relatively __52: few or many?__ class intervals. But no amount of clever manipulation can make relatively few cases tell the story of a large number.

49.

50.

51.

52.

The Graph is a Myth

A given collection of scores can be grouped in _53: only one way or different ways?_ , as you learned in the previous chapter. Each different way of grouping the scores will yield a different graph (whether that graph is a histogram, a _54_ polygon, or a _55_ percentage curve). Furthermore, given a particular way of grouping scores, the corresponding graph can be made squat or slender depending on the scale of one axis relative to the scale of the other axis. There is thus no such thing as *the one* graph of a given collection of observations.

53.

54.

55.

The Three Basic Characteristics of a Frequency Distribution

A given collection of scores (a given distribution) has three basic characteristics:

1. Where on the scale of possible scores is the center of the distribution located? Around what point on this scale do the scores cluster? What is the overall level of the scores? These questions deal with the _56_ tendency of the distribution.

56.

2. Do the scores cluster closely about their central point, or do they spread out considerably? To what degree do they spread out? These questions concern the _57_ of the distribution.

57.

3. Do the bulk of the scores fall in the middle of the distribution? Do the scores tend to cluster at a medium high point? At a medium low point? These questions concern the _58_ of the distribution.

58.

Kinds of Shape

Statisticians have provided names for certain shapes that commonly occur when one collects observations and describes them by means of a histogram or frequency polygon. Figure 3.12 on p. 44 is a series of _59: histograms or frequency polygons?_ illustrating these common shapes. A shape of great importance in statistical inference is the normal distribution, which is _60_ -shaped. In the normal _61_ , the bulk of the scores occur in the _62_ of the distribution, with relatively few low values and relatively few high values. If the bulk of the scores are not right in the middle but rather somewhat lower or higher, the shape is called _63_ . An extreme version of a skewed distribution, with the majority of the scores at the lowest possible value, is the _64_ -shaped distribution. A distribution in which scores tend to cluster at two points along the scale of possible values is called _65_ . Another possibility is the rectangular shape.

59.

60.
61.
62.
63.

64.

65.

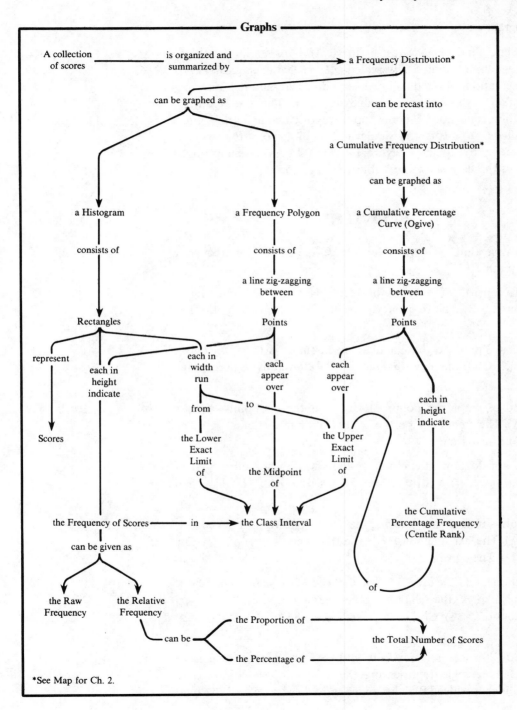

The Bar Diagram

Ch. 3, like Ch. 2 (and like the rest of the text), is primarily concerned with scores, which are numerically scaled 66: qualitative or quantitative?   66.
data. But observations of the qualitative kind do occur, and the frequency distribution that organizes and summarizes them, like the frequency distribution for a collection of scores, can be represented in a graph. Such a graph is called a 67 . It is very similar to the 68 , except that space is   67.
inserted between the rectangles. Figure 3.2 on p. 32 provides an example.   68.
It is a graph of the data shown in Table 2.3 on p. 15.

SELF-TEST

A histogram is a kind of: A) table.   B) centile rank.   C) frequency count.   D) graph.

1.   A   B   C   D

Which graph consists of a series of rectangles?   A) The cumulative percentage curve.   B) The frequency polygon.   C) The ogive.   D) The histogram.

2.   A   B   C   D

The graph most similar to the bar diagram in shape is the: A) ogive. B) histogram.   C) frequency   polygon.   D) cumulative   percentage curve.

3.   A   B   C   D

In which kind of graph are points plotted over the midpoints of class intervals?   A) The frequency polygon.   B) The bar diagram.   C) The ogive.   D) The histogram.

4.   A   B   C   D

The height of a histogram, in comparison to its width, should be: A) at least twice the size.   B) a little greater.   C) a little less.   D) less than half the size.

5.   A   B   C   D

In which graph is it necessary to bring the "curve" down to the X-axis at each end?   A) The bar diagram.   B) The frequency polygon.   C) The histogram.   D) The ogive.

6.   A   B   C   D

In a distribution of 132 scores, 75% of them fall below the value 84.5. Which symbols express the relationship between the figures 75 and 84.5 in this distribution?   A) $C_{132} = 75$.   B) $C_{132} = 84.5$.   C) $C_{75} = 84.5$. D) $C_{84.5} = 75$.

7.   A   B   C   D

A collection of 200 scores is carefully and correctly plotted in a large histogram. The rectangle representing the scores in the lowest class interval is 1.5 inches wide, 5.0 inches high, and takes up 25% of the total

8.   A   B   C   D

area of all the rectangles. How many scores fall within the lowest class interval?   A) 7.5   B) 25   C) 50   D) 75

Which is *not* one of the three basic characteristics of a frequency distribution?   A) Its shape.   B) Whether it is a sample or a population. C) Its variability.   D) Its central tendency.

9.   A   B   C   D

Here are four shapes that a frequency polygon may exhibit. Which is symmetrical with one "hump"?   A) Skewed.   B) **J**-shaped.   C) Normal.   D) Bimodal.

10.   A   B   C   D

CHAPTER 4

CENTRAL TENDENCY

OBJECTIVES

Ch. 3 informed you that a collection of scores has three fundamental properties: central tendency, variability, and shape. Ch. 4 describes three measures (indexes) of central tendency. In studying the chapter, work to understand these symbols, terms, concepts, principles, and techniques:

Purpose of a Measure of Central Tendency
Symbolism: X, Σ
Mode (Mo)
Median (Mdn)
  C_{50}
Arithmetic Mean ($\bar{X}$)
  Deviation Score • Balance Point of a Distribution • $\Sigma(X-\bar{X})=0$
Locations of Mode, Median, and Mean
  In Symmetrical Distributions
  In Skewed Distributions
    Positive Skew • Negative Skew
Properties of Mode, Median, and Mean
  Frequency of Usage • Mathematical Tractability • Sampling Stability • Effects of Extreme Scores and Skewness • Applicability to Open-Ended Distributions • Special Properties of the Mode

Also in studying Ch. 4, work to learn these skills:

Calculation of the Mode
Calculation of the Median
  For Small Samples of Ungrouped Scores
  For Large Samples of Grouped Scores
    Via the Graphic Method • Via the Arithmetic Method
Calculation of the Mean
  For Ungrouped Data • For Grouped Data

REVIEW

As you learned in the previous chapter, a collection of scores has three fundamental properties: central tendency, variability, and shape. This chapter is devoted to measures (indexes) of central tendency. A measure of central tendency is also known as a(n) 1 , and its purpose is to describe the 2 of the scores on the scale of possible values.

1.

2.

The Mode

The simplest measure of central tendency is the mode. It is the score that occurred with the 3 frequency in the distribution we are describing. If only a *grouped* frequency distribution is available, we cannot tell exactly how many times each individual score occurred, and in this case, the mode is taken to be the midpoint of the class interval with the greatest frequency.

3.

   The symbol for the mode is 4 .

4.

   A given distribution can have 5: only or more than? one mode.

5.

   The mode is largely insensitive to the exact values of the scores in the distribution. That is, many scores could change in many ways without any change in the mode. Some scores could become drastically higher or lower, but so long as the score that occurred the most often remains what it was, the mode would not change.

   To locate the mode in a distribution represented by a frequency polygon, find the highest point on the curve. The mode is the value down on the X axis directly below this highest point. See Figure 4.4 on p. 57 for two examples.

The Median

A second measure of central 6 is the median, which is the point along the scale of scores below which 7 of the cases fall. In the symbols of the centile system, the median is defined as 8 .

6.

7.

8.

   For a small sample of ungrouped scores, the 9 is calculated informally. If there are an odd number of scores, the median is taken to be the score in the 10 when the scores are arranged in order from low to high. But if there are an even number of scores, there will be no single score in the middle when they are arranged in order of magnitude. There will be two scores in the middle, and in this case, the median is taken to be the point 11 way between them.

9.

10.

11.

   The symbol for the median is 12 .

12.

A given distribution can have 13: only or more than? one median.   13.

The median is more sensitive than the mode to the exact values of the scores in the distribution, but it is not fully sensitive. The median responds to how many scores lie above or below it, but it does not respond to 14 from it the scores may be.   14.

In a frequency polygon, the median is the point along the X axis with the following property: When a straight line is erected up from this point, the line divides the area under the frequency polygon in 15 . The half of   15.
the area to the left of the median represents the half of the scores with the lowest values; the half of the area to the 16 of the median   16.
represents the half of the scores with the 17 values. (In general,   17.
proportion of 18 equals relative 19 of cases, as Section 3.8 of your   18.
text explained.)   19.

The Arithmetic Mean

A third measure of 20 tendency is the 21 mean, which is the 22 of   20.
all the scores divided by the 23 of scores. In ordinary speech, this   21.
measure is usually called the average, but to speak of "*the* average" is   22.
misleading, because there is more than one kind of average. It is best to   23.
avoid the term *average* altogether and to use the specific term *mode*, *median*, or *arithmetic mean* — except that the arithmetic mean is usually called simply the mean.

If the scores are available only as a grouped frequency distribution, to calculate the mean, let each score be represented by the 24 of the class   24.
interval in which it falls.

In statistical work, the symbol 25 represents a collection of scores   25.
(whether it is a sample or a population).* The mean of a sample symbolized in this way is then designated 26 ("eks bar"). The formula   26.
that defines the mean of a sample is 27 . In this formula, the symbols   27.
ΣX stand for the 28 of all the scores, and 29 stands for the number of   28.
scores. (ΣX should be read "the sum of eks.")   29.

A given distribution can have 30: only or more than? one mean.   30.

The mean 31: is or is not? sensitive to the exact value of every score in   31.
the distribution. You can understand its sensitivity if you imagine the scores to be weights spread along a weightless plank, as in Figure 4.3 on p. 56. In this physical analogy, the mean is the 32 of the distribution.   32.
That is, if the plank were supported only at this point, in the way that a see-saw rests on a crossbar, the plank would be perfectly level. But if one

*Be sure the symbol you write is the capital version of the letter.

of the weights on the plank were shifted, the balance point would change. Analogously, if any one of the scores in the distribution were changed in value, the __33__ would change. Thus the mean is responsive to the exact value (position) of each score.

33.

The fact that the mean is analogous to the balance point has to do with the amounts by which the scores deviate from the mean. Look at Figure 4.3 to review how these deviations are calculated. From each score the mean is subtracted. Scores below the mean have __34: positive or negative?__ deviations; any score at the mean has a deviation of zero; and scores above the mean have __35: positive or negative?__ deviations. If we do the arithmetic correctly, for any distribution, the sum of the negative deviations will exactly equal the sum of the __36__ (except that the sum of the negative deviations is a negative number). Thus the total of all deviations from the mean is __37__. The principle that the deviations from the mean sum to zero is expressed symbolically by the statement __38__.

34.

35.

36.

37.
38.

Relative Locations of the Mode, Median, and Mean

Where the mode, median, and mean lie with respect to each other depends on the shape of the distribution. If the distribution is symmetrical, the mean and the __39__ will always take the same value. If the distribution is not only symmetrical but unimodal (if the distribution has only one mode), the __40__ will also take this value. Look at the bottom of Figure 4.4 on p. 57 for an example. But if the distribution is skewed, the median is "pulled out" toward the "tail," away from the bulk of the scores, and the __41__ is pulled out even farther. Look at the top of Figure 4.4 for examples.

39.

40.

41.

Other Properties of the Mode, Median, and Mean

The measure of central tendency that is most commonly employed in statistical work is the __42__. It is used far more frequently than the others, because it has two highly desirable properties. First, it is the easiest to work with when there is statistical work to be done beyond simply describing the scores on hand (and usually there is such work: inferential statistics). Second, under ordinary circumstances, the mean is the __43: most or least?__ resistant to chance sampling __44__.

42.

On the other hand, for purposes of description, the mean is not always the best choice. As noted above, extreme scores (scores that are very much lower or higher than the others in a distribution) pull the mean __45: toward or away from?__ themselves. Thus in a highly __46__ distribution,

43.
44.

45.

46.

the mean is drawn well away from the bulk of the scores, and in such a
distribution, the __47__ is a better measure of what is typical.   47.

Another case in which the mean is not best for descriptive purposes
occurs when some scores are not known. Such a case arises when the best
we can say is that a score is, for example, "more than 5 min." or "greater
than 75 lb." A distribution with such scores is called __48__-ended, and it is   48.
not possible to calculate the mean without making some assumptions
about the unknown values. But if there are not too many missing figures,
we can calculate the median, and that is the best measure of __49__ here.   49.

What to Do for Qualitative Data

If the observations are not scores but qualitative data (which are classifi-
cations such as "blue" for eye color), there is only one measure of central
tendency that we can find. This is the __50__.   50.

CATCH YOUR OWN ERRORS

The mean of a collection of scores is like the balance point of a collection of weights
distributed along a plank, you have now learned. (The details are in Section 4.4.) On
the basis of this analogy, you can understand an important property of the mean
that is not cited in your text: *If there is any variability in a distribution of scores, the
mean must fall somewhere between the smallest score and the largest score.* Why?
Because the distribution of weights must balance somewhere between the leftmost
weight and the rightmost one.

This property of the mean permits you to tell quickly whether you have reached a
reasonable answer when you have computed a mean. Does your answer fall
somewhere between the smallest score and the largest one? If so, your answer could
be the mean. If not, you definitely made a mistake.

Getting an answer between the smallest and the largest scores does not guarantee
that you have correctly computed the mean, but it's still comforting to know the
answer lies in this interval. Getting an answer outside the limits of the distribution
does rule that answer out as the mean, of course.

It will be helpful to get into the habit of checking yourself by applying this
principle every time you calculate a mean. To practice, work these five very quick
exercises:

1. A distribution of scores runs from 19 to 43. Can the mean be 17.6?
2. A distribution of scores runs from 168 to 297. Can the mean be 208.8?
3. A distribution of scores runs from 3 to 19. Can the mean be 22.5?
4. The mean of a distribution is 44.6. Can the lowest score be 58?
5. The mean of a distribution is 224.9. Can the highest score be 512?

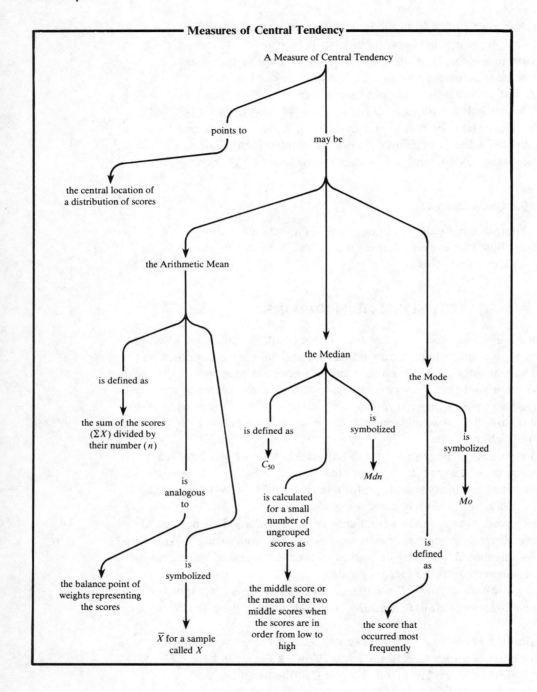

Measures of Central Tendency

A Measure of Central Tendency

points to

may be

the central location of
a distribution of scores

the Arithmetic Mean

the Median

the Mode

is defined as

is defined as

is
symbolized

is
symbolized

the sum of the scores
(ΣX) divided by
their number (n)

C_{50}

Mdn

Mo

is
analogous
to

is calculated
for a small
number of
ungrouped
scores as

is
defined
as

the balance point of
weights representing
the scores

is
symbolized

the middle score or
the mean of the two
middle scores when
the scores are in
order from low to
high

the score that
occurred most
frequently

$\overline{X}$ for a sample
called X

There is a general lesson here: *When carrying out a calculation in statistics, ask yourself, "What is the quantity that I am trying to determine?" Then on the basis of what you know about this quantity, ask yourself whether your answer is a sensible value for the quantity.*

In the present instance, when calculating $\Sigma X/n$, ask yourself, "Just what is this quantity $\Sigma X/n$?" The answer, as Ch. 4 informed you, is the arithmetic mean of the distribution of scores. On the basis of what you've just learned about the mean, you can then ask, "Is my answer somewhere between the smallest score in the distribution and the largest one?" If yes, your answer is reasonable (but it still needs to be verified). If no, you have discovered instantly that you made an error.

SELF-TEST

An average is a measure of: A) frequency. B) sample size. C) shape. D) central tendency.

1.   A B C D

Which measure, if any, can take two values for a given distribution? A) The mode. B) The median. C) The mean. D) None of these.

2.   A B C D

The median is sensitive to: A) the exact value of every score. B) only the highest and the lowest score. C) the number of scores that lie above it and below it. D) how far away from it the scores lie.

3.   A B C D

A score below the mean of its distribution has a deviation from the mean that is: A) positive. B) zero. C) negative. D) impossible to compute.

4.   A B C D

The sum of the positive deviations from the mean is always: A) a negative number. B) zero. C) the same as the sum of the negative deviations from the mean (except for whether the sum is negative or positive). D) the balance point of the distribution.

5.   A B C D

If the one topmost score in a distribution is doubled, which measure will change? A) The mode. B) The median. C) The mean. D) All three.

6.   A B C D

In a distribution with a symmetrical shape, which two measures will always take the same value? A) The mode and the median. B) The mode and the mean. C) The median and the mean. D) The median and the number of scores.

7.   A B C D

If the mode of a distribution is 10, the median is 15, and the mean is 20, what is the probable shape of the distribution? A) Skewed left. B) Skewed right. C) Normal. D) Rectangular.

8.   A B C D

Suppose a research team classifies their subjects as straight-haired, wavy-haired, curly-haired, or bald. Which is the best measure of central tendency for their observations? A) The mode. B) The median. C) The mean. D) The mode and the median are equally good.

9.   A   B   C   D

Why is it impossible to compute the mean of an open-ended distribution (without making assumptions about the values of the unknown scores)? A) The mean depends on the number of scores, and the number of scores is not known. B) The mean depends on the score that occurred the most frequently, and the score that occurred the most frequently is not known. C) The mean depends on the fiftieth centile, and the fiftieth centile is not known. D) The mean depends on the total value of all the scores, and the total value is not known.

10.   A   B   C   D

CHAPTER 5

VARIABILITY

OBJECTIVES

In Ch. 3, you learned that every distribution of scores has three basic properties: central tendency, variability, and shape. Ch. 4 introduced three measures of central tendency. Ch. 5 now describes four measures of variability. In studying the chapter, work to understand the following symbols, terms, concepts, principles, and techniques:

Measures of Variability
    Purpose • As Supplements to Averages • Importance in Statistical
    Inference • What Measures of Variability Do Not Do
Range
Semiinterquartile Range (Q)
    Q_1 • Q_2 • Q_3 • C_{25} • C_{50} • C_{75}
Deviation Score (x)
    $\Sigma x = 0$ • The Purpose of Squaring Deviation Scores
Variance (S^2)
    Disadvantage as a Descriptive Measure
Standard Deviation (S)
    Incorporation of the Mean
    Calculation of Σx^2 from Raw Scores
        ΣX^2 • $(\Sigma X)^2$
    Calculation for Grouped Data
Comparison of Measures of Variability
    Frequency of Usage • Mathematical Tractability • Sampling Stability • Responsiveness to Total Condition of the Distribution • Effect of Extreme Scores and Skewness • Coordination with Measures of Central Tendency • Applicability to Open-Ended Distributions • Dependence of the Range on Sample Size
Normal Distribution
    The Standard Deviation as a Measure of Distance from the Mean
    Percentage of Cases in Intervals $\overline{X} \pm 1S$, $\overline{X} \pm 2S$, $\overline{X} \pm 3S$
Comparing Means of Two Distributions

Also in studying Ch. 5, work to learn these skills:

Calculation of the Range
Calculation of the Semiinterquartile Range via the Graphic Method
Calculation of Deviation Scores
Calculation of the Variance
    From Deviation Scores • From Raw Scores
Calculation of the Standard Deviation
    From Deviation Scores • From Raw Scores

REVIEW

The Purpose and Nature of Measures of Variability

The purpose of a measure of _1_ is to indicate the extent to which the scores in a distribution scatter about or cluster together. A proper description of a distribution often requires not just a measure of central _2_ but also a measure of _3_. And measures of variability are particularly important in statistical inference, where we are concerned with the effects of _4_ factors on the results of sampling.

A measure of variability is a summary figure that describes the spread of an entire collection of scores. It _5: specifies or does not specify?_ how far any particular score diverges from the center of the collection. It _6: provides or does not provide?_ information about the level of the scores. It gives _7: some or no?_ clue to the shape of the distribution.

1.
2.
3.
4.

5.

6.
7.

The Range

The simplest measure of variability is the range, which is the _8_ between the highest and the _9_ score.

There is no symbol for the range in common use.

8.
9.

The Semiinterquartile Range

Every distribution has three quartile points. These are the three points along the scale of scores that divide the distribution into _10: how many?_ parts, each containing _11: what fraction?_ of the cases. These three points are symbolized _12_, respectively. In terms of the centile system, _13_ is the same as C_{25}; Q_2 is the same as _14_; and Q_3 is the same as _15_. (Q_2, or C_{50}, is one of the measures of central tendency, the _16_.)

10.
11.
12.
13.
14.
15.
16.

The semiinterquartile range is a measure of 17 constructed from the   17.
first and 18 quartiles. The distance between these two quartiles, $Q_3 -$ 18.
Q_1, is the range of the middle 19 % of the scores; we could thus call it   19.
the interquartile range. The *semi*interquartile range is half of this quantity. In symbols, to compute the semiinterquartile range, calculate 20 .   20.

The symbol for the semiinterquartile range itself is 21 .   21.

Be sure not to confuse Q with Q_1, Q_2, or Q_3. Q is a 22: distance   22.
or point? , while Q_1, Q_2, and Q_3 are 23: distances or points? .   23.

Deviation Scores

The other two measures of variability are based on the deviation scores
that correspond to the raw scores in a distribution. The symbol for a
deviation score is 24 * (read "little eks"). In terms of this symbol and   24.
others already introduced, the equation that defines a deviation score is
 25 .   25.

A deviation scores states the position of the corresponding raw score
relative to the 26 of the distribution. The sign of the deviation score,   26.
plus or minus, indicates whether the raw score is above or below the 27 ,   27.
respectively. The magnitude of the deviation score (the number itself,
ignoring the sign), indicates the distance between the raw score and the
 28 . For example, in a distribution with a mean of 10, a raw score of 5   28.
has a deviation score of 29 . The deviation score, x, is calculated as   29.
$X - \overline{X}$, which here is $5 - 10 = -5$. In the deviation score of -5, the
minus sign indicates that the raw score is 30 the mean, and the 5   30.
indicates that it is 31 units distant from the 32 .   31.

If all the raw scores in a distribution are converted to deviation scores,   32.
the collection of deviation scores indicates how much variability there is
among the raw scores. The more the raw scores spread out, the
 33: farther from or closer to? their own mean they are. And the farther   33.
from their own mean they are, the 34: larger or smaller? the deviation   34.
scores (ignoring the minus signs) will be. The other two measures of
variability are both based on the 35 scores.   35.

The Variance

The variance is the 36 of the squared 37 scores. If the raw scores   36.
are widely scattered among themselves and thus have deviation scores   37.
that are large (large positive and large negative), the squared deviation

---

*Be sure you make the lower-case version of the letter. I suggest you draw it not merely
small but with a hook, like this: ✘

scores will also be large, and the mean of the squared deviation scores will be too. In this way, the variance reflects the variability among the raw scores.

The symbol for the variance of a sample is __38__. The formula that defines the variance, in terms of the deviation scores, x, is __39__. In terms of the raw scores, the formula that defines the variance is __40__.

Note that any expression of the form Σ/n is a mean. In Ch. 4 we encountered $\Sigma X/n$, which defines the mean of the raw scores, X. Now we have $\Sigma x^2/n$, or $\Sigma(X - \bar{X})^2/n$, which defines the mean of the squared deviation scores, $x^2 = (X - \bar{X})^2$. (The mean of the squared deviation scores is not called a mean, though; it is called the __41__)

The variance __42: is or is not?__ useful in descriptive statistics, because it is expressed in squared units of __43__. (For example, if the scores in a distribution are the heights of human subjects, the mean will also be a height—68.3 inches, perhaps—but the variance will be a squared height such as 5.3 squared inches. And the concept of squared inches makes no sense for descriptive purposes here.) The variance is, however, useful in __44__ statistics.

38.
39.
40.

41.

42.
43.

44.

The Standard Deviation

The standard deviation is the __45__ of the __46__. By taking the square root of the variance, we return to a measure that is in the same units as those of the raw scores and is thus more useful for purposes of __47: description or inference?__.

The symbol for the standard deviation of a sample is __48__, and the formula that defines it, in terms of the deviation scores, is __49__.

If the scores are available only as a grouped frequency distribution, it __50: is or is not?__ possible to use this formula. In this case, let each score be represented by the __51__ of the class interval in which it falls. (This is the same strategy used to calculate the mean in such a case.)

45.
46.

47.

48.
49.

50.
51.

Comparison of the Measures of Variability

The measure of variability that is most commonly used for descriptive purposes is the __52__. The standard deviation is superior to the variance as a descriptive measure for the reason explained just above. The standard deviation is superior to the range and to the semiinterquartile range for these three reasons: First, the standard deviation is easy to use in further

52.

calculations, whereas the range and the __53__ are not. Second, the standard deviation is the __54: most or least?__ resistant to sampling __55__. Third, the standard deviation is responsive to the exact __56__ of every score in the distribution, whereas the __57__ and the semiinterquartile range are not.

53.
54.
55.
56.

The fact that the standard deviation is responsive to the exact __58__ of every score is sometimes a disadvantage, however. If a distribution contains even a very few extreme scores (scores extremely far above or far below the rest), the standard deviation (and the range) will be inflated and will give a misleading impression of the variability of the distribution. For such a distribution, the best measure of the variability is the __59__.

57.
58.

59.

The standard deviation is also a poor choice for an open-ended distribution—in fact, it is impossible to calculate the standard deviation (or the range) without making assumptions about the values of the missing scores. In such a case the only reasonable alternative is the __60__.

60.

Opposite to the standard deviation in usefulness is the range. It __61: is or is not?__ sensitive to the exact nature of the distribution, because it is wholly determined by just __62: how many?__ scores—the highest and the __63__. It tells us nothing about what happens in between them. Furthermore, a single errant score, either high or low, can cause the range to give a grossly inflated impression of the variability in the distribution. The range is also the __64: most or least?__ resistant to __65__ variation, and in many types of distribution, the range depends on the sample __66__. The larger the sample, the __67: larger or smaller?__ the range is likely to be. In contrast, the other measures do not systematically change as the sample size changes.

61.
62.
63.

64.
65.
66.
67.

The Standard Deviation and the Normal Distribution

If a collection of scores (sample or population) has the particular bell shape that is called normal, the span from one standard deviation below the __68__ to one standard deviation above it contains about __69__ % of the scores. The span from two __70__ below the mean to two above contains about __71__ % of the scores. And the span from three __72__ below the mean to three above contains about __73__ % of the scores.

68.
69.
70.
71.
72.
73.

In general, even if a distribution departs considerably from normality in its shape, the span $\overline{X} \pm 1S$ will include __74: a minority or the majority?__ of the scores. The span $\overline{X} \pm 2S$ will include the great majority and sometimes even all the scores. The span __75__ will include all or all but a very few cases.

74.
75.

Comparing Means of Two Distributions

Research often involves a comparison of the __76__ of one group with
the __77__ of another group. The size of the difference between the means
usually has no meaning in itself. To provide a frame of reference by
which to judge whether a difference is large or small, the __78__ is useful.
Divide the difference between the means by the __79__. (If the standard
deviations of the two groups differ, use the mean of the standard
deviations as the divisor.) A difference of .1 standard deviation is
negligible, while a difference of __80__ standard deviation is of some
importance.

76.

77.

78.

79.

80.

SPECIAL HELP WITH THE VARIANCE AND THE STANDARD DEVIATION

To understand how the variance and the standard deviation function as measures
of the variability among the scores in a distribution, remember that they are based
on the deviations between the scores and the mean.

The mean will always fall somewhere between the lowest score in the distribution
and the highest one, as the previous chapter of this *Guide* noted, so some scores will
be below the mean, and some will be above it. For an illustration, look at the
left-most column in the table below. This is not a frequency distribution, just a
listing of some scores in order from high down to low. Check the computation of the
mean, and note that it falls between the low of 11 and the high of 23.

Raw Score, X	Deviation Score, x	Squared Deviation Score, x^2
23	+5	25
22	+4	16
19	+1	1
17	−1	1
16	−2	4
11	−7	49

$\overline{X}-$ (line between 19 and 17)

$\Sigma X = 108$ $\Sigma x = 0$ $\Sigma x^2 = 96$

$n = 6$ $S^2 = 96/6$

$\overline{X} = \Sigma X/n$ $= 16.0$

$= 108/6 = 18.0$ $S = \sqrt{16.0} = 4.0$

Where, now, are the raw scores in relation to the mean? The deviation scores in
the second column tell you. Raw scores below the mean have negative deviations;
raw scores above the mean have positive deviations; scores right at the mean would

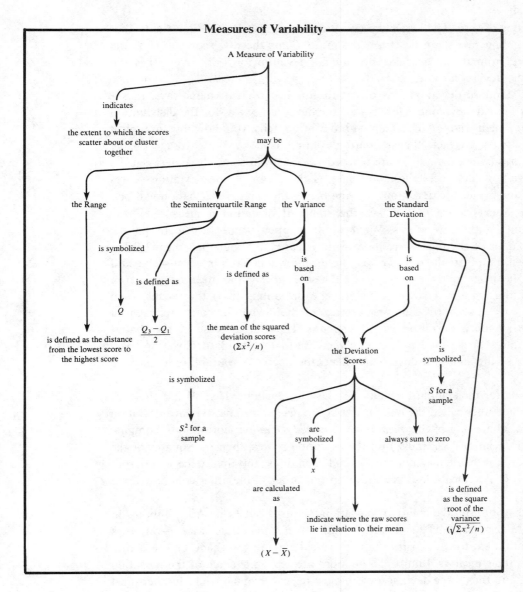

Measures of Variability

A Measure of Variability

indicates

the extent to which the scores scatter about or cluster together

may be

the Range

the Semiinterquartile Range

the Variance

the Standard Deviation

is symbolized

is defined as

is defined as

is based on

is based on

Q

is defined as the distance from the lowest score to the highest score

$\dfrac{Q_3 - Q_1}{2}$

the mean of the squared deviation scores $(\Sigma x^2 / n)$

the Deviation Scores

is symbolized

is symbolized

S^2 for a sample

are symbolized

always sum to zero

S for a sample

x

are calculated as

indicate where the raw scores lie in relation to their mean

is defined as the square root of the variance $(\sqrt{\Sigma x^2 / n})$

$(X - \bar{X})$

have deviations of zero. The larger the deviation score, ignoring its sign, the farther the corresponding raw score lies from the mean. Thus the raw score of 11 is the farthest score from the mean here, because its deviation of 7 (really −7) is the greatest of the deviation scores.

Now: the standard deviation of a distribution is just what its name says. It is a standard, or typical, deviation. How big do the deviation scores for the distribution run (ignoring their signs)? The standard deviation tells you how big a typical deviation score is. In fact, the standard deviation is a kind of average of the deviation scores (when their signs are ignored). It must fall somewhere between the smallest deviation and the largest one, just as the mean of the deviation scores would. And the bigger the deviations are, the bigger the standard deviation will be.

It is this latter feature that makes the standard deviation a measure of the variability among the raw scores. What would happen if the raw scores in the left-hand column of the table were more widely scattered about their mean? What would happen, that is, if the scores *above* the mean became even larger in value and thus moved farther above the mean, while the scores *below* the mean became even smaller and thus moved farther below it? The deviations from the mean would increase, of course, and so would any average of the deviation scores. As such an average, the standard deviation would increase. Thus an increase in the scatter among the raw scores would produce an increase in the standard deviation. And thus the size of the standard deviation reflects the degree of variability among the scores.

The formula for the standard deviation is easy to remember if you learn that this quantity is sometimes called the *root mean squared deviation*. The standard deviation is indeed the square *root* of the *mean* of the *squared deviation* scores. So to compute the standard deviation, you must find the deviation scores, then the squares of the deviation scores, then the mean of the squared deviations, and finally the square root of this mean. These calculations are shown in this order in the other columns of the table above.

For this distribution, the standard deviation, *S*, turns out to be 4.0. Think again what this number means. The standard deviation *is* a *standard* deviation—that is, a typical deviation. So for this distribution, a typical amount by which the raw scores deviate from their mean is 4 units. To be more precise, an average of the deviation scores is 4.0, and thus some deviation score must be less than 4.0 while another must be more than 4.0. Look to see how these things are true in the table. The deviation scores in the second column (ignoring their signs) run from 7 and 5 down through 4 and 2 to 1. The value for the standard deviation, 4.0, does indeed fall between the 7 and the 1, and it is a pretty good average for those numbers.

Note that the standard deviation is calculated by doing two operations and then undoing them—which gets you back to something comparable to what you began with. You start with deviation scores, and you do two things to them: square them, and then sum the squares. Then you undo those things: you divide by the number of cases, which undoes the summing; and you take the square root, which undoes the squaring. The result, the standard deviation, is comparable to the deviation scores with which you began: it *is* a standard, or typical, deviation.

Look again now at the computations in the table above. On the way to the standard deviation in the lower right corner, the variance, S^2, turned up, for the variance is nothing more than the mean of the squared deviation scores. Remembering that it's a mean will help you tell if you have a reasonable value when you've computed a variance. As the mean of the squared deviation scores, it must fall somewhere between the largest squared deviation and the smallest one. In the table, observe that 16.0 does indeed fall between the high of 49 and the low of 1.

Exercises

To practice conceptualizing the variance and the standard deviation in the terms introduced above, fill in the tables below, which are like the example you just examined. As you fill in the tables, *think what you are doing*. Here's a list of relevant thoughts that are well worth thinking.

1. The **mean** ($\overline{X}$) is an average of the raw scores and thus indicates about how large they run. It takes an intermediate value, falling somewhere between the smallest raw score and the largest. It serves as the reference point for the deviation scores (the x's).

2. The **deviation scores** (x's) indicate where the corresponding raw scores (the x's) lie in the relation to the mean. A deviation score indicates whether the associated raw score is above ($+$) or below ($-$) the mean, and how far from the mean the raw score lies. Taking their signs into account, $\Sigma x = 0$, as p. 56 of the text informed you.

3. The **variance** (S^2) is the mean of the squared deviation scores (the mean of the x^2's). As an average of the squared deviation scores, it indicates about how large they run, and it takes an intermediate value, somewhere between the smallest and the largest squared deviation.

4. The **standard deviation** (S) is a typical value for the deviation scores (ignoring their signs). It serves as an average for the deviation scores (x's), indicating about how large they run. It takes an intermediate value, somewhere between the smallest deviation and the largest. It is calculated by doing two operations (squaring the deviation scores and summing them) and then undoing the operations (dividing the sum by n and taking the square root of the results). The undoing gets you back to something comparable to what you started with, which was the deviation scores.

Raw Score, X	Deviation Score, x	Squared Deviation Score, x^2
13		
12		
9		
7		
6		
1		

$\Sigma X =$ $\Sigma x =$ $\Sigma x^2 =$

$n =$ $S^2 = \quad /$

$\overline{X} = \Sigma X / n$ $=$

$= \quad / \quad =$ $S = \sqrt{\quad} =$

Your standard deviation for the little distribution in that table should be 4.0, the same as the value for the example first offered. Why do the two distributions have the same standard deviation? Yes, it's because they have the same degree of variability, but why, *exactly*, do the two standard deviations work out to be the same number? (The answer is in the back of this *Guide*, along with the answers to the other questions in these exercises.)

Raw Score, X	Deviation Score, x	Squared Deviation Score, x^2
14		
13		
10		
10		
8		
5		

$\Sigma X =$ $\Sigma x =$ $\Sigma x^2 =$

$n =$ $S^2 = \quad /$

$\overline{X} = \Sigma X / n$ $=$

$= \quad / \quad =$ $S = \sqrt{\quad} =$

As you fill in these tables, don't work like a mindless robot. Every number you write has some meaning, in accord with the principles numbered 1 - 4 above. Think through what you're doing in terms of those principles.

Raw Score, X	Deviation Score, x	Squared Deviation Score, x^2
14		
9		
9		
9		
9		

$\Sigma X =$
$n =$
$\overline{X} = \Sigma X / n$
$\quad = \quad / \quad =$

$\Sigma x =$

$\Sigma x^2 =$
$S^2 = \quad /$
$\quad =$
$S = \sqrt{\quad} =$

Raw Score, X	Deviation Score, x	Squared Deviation Score, x^2
7		
7		
7		
5		
4		
4		
3		
3		
3		
2		
2		
1		

$\Sigma X =$
$n =$
$\overline{X} = \Sigma X / n$
$\quad = \quad / \quad =$

$\Sigma x =$

$\Sigma x^2 =$
$S^2 = \quad /$
$\quad =$
$S = \sqrt{\quad} =$

More Exercises

In real life, computing variances and standard deviations from deviation scores—the way you've just been doing it—is usually laborious, because the mean of a distribution is typically some absurdly unwhole number like 15.632, and the deviations are then also decimal fractions. Fortunately, there are ways to compute the right answers without fussing with deviation scores.

The alternative techniques require only the raw scores (the X's). The basic formula is:

$$\Sigma x^2 = \Sigma X^2 - (\Sigma X)^2/n$$

(Read this as "The sum of little-x-squared equals the sum of big-X-squared minus the-sum-of-big-X-the-quantity-squared over n.") Thus whenever you need Σx^2, you can substitute the expression on the right of the equals sign. To practice using this formula and to see that it is trustworthy, use the tables below to recompute Σx^2 for each of the distributions you worked with above. Then look to see if the result is the same as what you previously calculated for Σx^2. In the tables below, the raw scores are relisted. The first one shows the computations for the data in the table used as the initial example.

Squared RAW Score, X^2*	Raw Score, X	
529	23	
484	22	$\Sigma x^2 = \Sigma X^2 - (\Sigma X)^2/n$
361	19	$= 2040 - (108)^2/6$
289	17	$= 2040 - 11664/6$
256	16	$= 2040 - 1944$
121	11	$= 96 =$ the value computed
$\Sigma X^2 = 2040$	$\Sigma X = 108 \quad n = 6$	directly in the first table

*To make it clear that X^2 differs from x^2, the column of X^2 values is listed on the left side of the raw scores.

Squared RAW Score, X^2	Raw Score, X	
	13	$\Sigma x^2 = \Sigma X^2 - (\Sigma X)^2/n$
	12	$= \quad -(\quad)^2/$
	9	$= \quad - \quad /$
	7	$= \quad -$
	6	$=$ should $=$ the value
	1	computed directly
$\Sigma X^2 =$	$\Sigma X = \quad n =$	in the second table

Squared RAW Score, X^2	Raw Score, X	
	14	$\Sigma x^2 = \Sigma X^2 - (\Sigma X)^2/n$
	13	$= \quad -(\quad)^2/$
	10	$= \quad - \quad /$
	10	$= \quad -$
	8	$= \quad$ should $=$ the value
	5	computed directly
$\Sigma X^2 =$	$\Sigma X = \quad n =$	in the third table

Squared RAW Score, X^2	Raw Score, X	
	14	$\Sigma x^2 = \Sigma X^2 - (\Sigma X)^2/n$
	9	$= \quad -(\quad)^2/$
	9	$= \quad - \quad /$
	9	$= \quad -$
	9	$= \quad$ should $=$ the value
$\Sigma X^2 =$	$\Sigma X = \quad n =$	computed directly
		in the fourth table

Squared RAW Score, X^2	Raw Score, X	
	7	$\Sigma x^2 = \Sigma X^2 - (\Sigma X)^2/n$
	7	$= \quad -(\quad)^2/$
	7	$= \quad - \quad /$
	5	$= \quad -$
	4	$= \quad$ should $=$ the value
	4	computed directly
	3	in the fifth table
	3	
	3	
	2	
	2	
	1	
$\Sigma X^2 =$	$\Sigma X = \quad n =$	

Mnemonic Tip

That formula you practiced using in the exercise above is worth memorizing. Write it out several times, taking care to distinguish x from X and ΣX^2 from $(\Sigma X)^2$. Note again that the formula is read "The sum of little-x-squared equals the sum of big-X-squared minus the-sum-of-big-X-the-quantity-squared over n." Rehearse these words as well as the equation they describe.

DESCRIPTIVE STATISTICS IN USE

In 1973, the U. S. Consumer Product Safety Commission received reports indicating that spray adhesives damage chromosomes and can cause a pregnant woman to bear a deformed child. The Commission banned the sale of such adhesives in August of that year and warned all persons who had come into contact with them, especially pregnant women, to consult a physician for a chromosome study.

To assess the impact of the warning, two researchers subsequently mailed a questionnaire to all the Americans listed in a directory of professionals who do diagnostic cytogenetics (the study of the cellular basis of heredity) and genetic counseling. Only five of these individuals failed to respond. The questionnaire asked for the number of persons who had consulted the respondent about exposure to spray adhesives and the number whose chromosomes the respondent had actually studied. The researchers reported the following data.

It may appear from these tables that the distributions were open-ended, but except for the respondents who reported merely "some inquiries" or "some studies," the researchers knew all values of the variables of interest.

Number of People Consulting Respondent about Exposure to Spray Adhesive	f
None	52
1– 5	68
6–10	31
11–15	8
16–20	8
21–25	2
Over 25	7
	$\overline{176}$*

Range 0–200. $\bar{X} = 6.81$ inquiries per respondent. $C_{25} = 0$, $C_{50} = 3.2$, $C_{75} = 7.6$.

*Six additional respondents reported "some" inquiries.

Number of Persons Whose Chromosomes were Studied by Respondent	f
None	49
1– 5	58
6–10	13
11–15	4
16–20	2
21–25	1
Over 25	1
	$\overline{128^*}$

Range 0–44. $\overline{X} = 2.97$ studies per respondent. $C_{25} = 0$, $C_{50} = 2.0$, $C_{75} = 4.2$.

*Two additional respondents reported "some" studies.

As you can see, the researchers did not follow standard practice in constructing their grouped frequency distributions. Nevertheless, these data illustrate a number of techniques of descriptive statistics and offer you a chance to review much of what you've learned thus far.

1. In what ways do those tables depart from the principles governing the construction of a grouped frequency distribution?

2. What is the shape of the first distribution?

3. What is the shape of the second distribution?

4. What is the median of the first distribution?

5. What is the median of the second distribution?

6. Why is the mean higher than the median in the first distribution?

7. Why is the mean higher than the median in the second distribution?

8. What is Q for the first distribution?

9. What is the Q for the second distribution?

10. The researchers reported the values of Q for their distributions, a rare occurrence in the scientific literature. Why is Q better than S as a measure of variability for these distributions?

11. For the 176 respondents who reported an exact number, what was the *total* number of people who consulted with them about the effects of exposure to spray adhesives?

12. For the 128 respondents who reported an exact number, what was the *total* number of people whose chromosomes were studied?

13. Why did the researchers employ only descriptive statistics with no inferential techniques?

Don't be upset if you tried computing C_{50} and C_{75} for the two distributions and failed to get the figures the researchers reported. Their figures do not correspond to the frequencies in the tables. They must have computed the quartiles from the ungrouped data.

Important note: The ban on spray adhesives was withdrawn in six months when the purported correlations between exposure to the chemicals and chromosomal damage or birth defects could not be confirmed, and no toxicity could be demonstrated for the adhesives. In fact, investigators who reexamined the slides that had initially been believed to show chromosome damage in exposed persons did not agree with the original interpretation.

Reference: E. B. Hook & K. M. Healy, "Consequences of a Nationwide Ban on Spray Adhesives Alleged to be Human Teratogens and Mutagens," *Science*, 1976, *191*, 566 - 567.

SELF-TEST

Measures of variability are important in: A) descriptive statistics.   1.   A B C
B) inferential statistics.   C) neither of these.   D) both of these.

What does a measure of variability do?   A) It specifies how far a   2.   A B C
particular score diverges from the center of its distribution.   B) It
describes the spread of the entire distribution.   C) It provides information about the overall level of the scores.   D) It gives a clue to the shape
of the distribution.

A measure of central tendency is a _____, and a measure of variability is a _____   A) distance ... distance.   B) distance ... location.   C) location ... distance.   D) location ... location.

3.   A   B   C   D

Another name for the median is: A) C_{50}.   B) Q_2.   C) both of these.   D) neither of these.

4.   A   B   C   D

In a distribution of 50 scores, the lowest is 10 and the highest is 40. The deviation score for the raw score of 10 is: A) -30.   B) -40.   C) zero.   D) impossible to compute without more information.

5.   A   B   C   D

Which measure of variability is *not* based on the deviation scores?   A) The semiinterquartile range.   B) The variance.   C) The standard deviation.   D) None; that is, all three measures named here *are* based on the deviation scores.

6.   A   B   C   D

The variance is the _____ of the _____ deviation scores: A) largest ... negative   B) smallest ... positive   C) mean ... raw   D) mean ... squared.

7.   A   B   C   D

What is the symbol for the standard deviation of a sample?   A) S B) S^2 C) x D) $\overline{X}$

8.   A   B   C   D

If a distribution contains a few extremely low scores, the standard deviation will be: A) misleadingly low.   B) an accurate descriptive measure of the variability of the distribution.   C) misleadingly high.   D) impossible to compute.

9.   A   B   C   D

The standard deviation of any reasonably representative sample of scores on a Scholastic Aptitude Test (SAT) is about 100 points. Suppose this year's entering class at your college has a mean SAT-Math score of 538, while last year's entering class had a mean SAT-Math score of 528. The difference between these two means is: A) negligible.   B) of some importance.   C) highly significant.   D) impossible to compute.

10.   A   B   C   D

CHAPTER 6

NORMAL DISTRIBUTIONS AND STANDARD SCORES

OBJECTIVES

Ch. 4 was concerned with central tendency, and Ch. 5 with variability. Ch. 6 is devoted to the third of the three basic aspects of a distribution of scores—its shape. The chapter introduces the normal curve, which provides a model for the shapes of some distributions. In studying the chapter, work to understand the following symbols, terms, concepts, principles, and techniques:

The Normal Curve
    As a Model for the Shape of a Distribution of Raw Scores   •   As a Model for the Shape of a Distribution of Sample Statistics   •   As a Family of Curves
z Score
    Translating a Raw Score to a z Score   •   Translating a z Score to a Raw Score   •   Shape of a Distribution of z Scores   •   Mean of a Distribution of z Scores   •   Standard Deviation of a Distribution of z Scores
Finding Areas Under the Normal Curve, Given One or Two Scores
Finding Scores in a Normal Distribution, Given an Area
Use of z Scores to Compare Raw Scores from Two Distributions
    Principle of z-Score Equivalence
Standard Scores other than z Scores
Centile Scores in a Normal Distribution
    Disadvantage   •   Advantage   •   Translation from a z Score   •   Translation to a z Score

Also in studying Ch. 6, work to learn these skills:

Calculation of a z Score, Given X, $\overline{X}$, and S
Calculation of a Raw Score, Given z, $\overline{X}$, and S
Use of a Table of Areas under the Normal Curve (Appendix F)
    Calculation of a Proportion of Cases, Given One or Two Scores, Raw or z
    Calculation of a Score, Raw or z, Given a Proportion of Cases

REVIEW

The Normal Curve as a Model for the Shape of a Distribution

The normal curve is a model—that is, a representation, a description. It represents just one feature of a collection of numbers—its 1 . Some distributions of scores that arise in research have a shape for which the normal curve is a poor model, but there are many whose shape is closely described by the normal 2 .

1.

2.

The normal curve also functions well as a model for the distribution of many sample 3 . Suppose, for example, that we draw a very great number of random samples from some population of scores. (All the samples should be of the same size.) For each sample, we compute the mean, which is a statistic characterizing the sample, and we cast these statistics into a distribution. The shape of the distribution of this large number of means would be close to that specified by the normal curve. The normal curve is thus valuable in both descriptive statistics and 4 statistics.

3.

4.

The normal curve is a model only of the shape of a distribution. It does not specify the central tendency of a distribution (it does not give the value of the mean, for example), nor does the model specify the variability (it does not give the value of the standard deviation, for example). Thus different distributions can all conform to the shape called 5 , even though the distributions differ in the mean, in the standard deviation, or in both.

5.

Look at Figure 6.1 on p. 90 for examples to illustrate this important point. Each curve in Figure 6.1 is a stylized (that is, smoothed or schematized) version of a 6: what kind of graph? . In what sense do all the distributions illustrated in Figure 6.1 have the same shape, the one called normal? The shapes are all the same in the sense that the area under the curve is distributed in the same way in each instance. Consider, for example, the span that runs from one standard deviation below the mean to one standard deviation above the mean. No matter what values the standard deviation and the mean take, if a distribution has a normal shape, over this interval there lies 7 % of the area under the curve, as Section 5.7 explained. Over the interval from two standard deviations below the mean to two standard deviations above it—no matter what the standard deviation and the mean are—there lies 8 % of the area. And over the interval from three standard deviations below the 9 to three standard deviations above it—again no matter what particular numbers are involved—there lies 10 % of the area.

6.

7.

8.

9.

10.

The percentages cited here are not merely proportions of area on a graph. They are also proportions of cases (scores). In general, the proportion of 11 in any part of a histogram or frequency 12 equals the proportion of cases with scores in the same location. Thus if a distribution of numbers—whether they are scores or sample statistics— has a shape that is normal, no matter what the mean and what the standard deviation might be, we know the following about how the cases are distributed relative to one another: 13 % of the cases fall in the interval $\overline{X} \pm 1S$; 14 % of the cases fall in the interval $\overline{X} \pm 2S$; and 99.7% of the cases fall in the interval 15 .

11.
12.

13.
14.
15.

Characteristics of the Normal Curve

Exactly what is the shape of the normal curve? It is a specific kind of bell shape. It is 16: symmetrical or asymmetrical? and 17: unimodal or bimodal? . Because it is symmetrical and unimodal, the mean, the median, and the 18 of any distribution with a normal shape all take the same value. Starting at the center of the curve and working outward in each direction, the height of the curve descends gradually at first, then faster, but then finally slower. (It is thus a particular kind of bell shape.)

In the mathematical abstraction that is the normal curve, although the curve descends promptly toward the 19: horizontal or vertical? axis, the curve never actually touches this axis, no matter how far down below the mean or how far up above the mean one goes. You might then think that if the curve never touches the 20: X or Y? axis, a large proportion of the total area under the curve must lie in the extremes, but this is not so. As noted above, 99.7% of the area, representing 21 % of the cases in any distribution whose shape is normal, lies within the interval 22 .

16.
17.
18.

19.

20.

21.
22.

z Scores

If a large number of people representative of the general population take a certain kind of IQ test, the Wechsler Adult Intelligence Scale (the WAIS), the distribution of their scores will have a mean about 100 and a standard deviation about 15. If a large number of people representative of the population of college students who aspire to graduate school take the Graduate Record Exam (the GRE), the distribution of their scores will have a mean about 500 and a standard deviation about 100 (on each of the three parts of the aptitude test). Given these states of affairs, there is an important way in which an IQ score of 100 from the first distribution corresponds to a GRE score of 500 from the second distribution: each falls at the mean of its distribution. There is also an important way in which an IQ score of 115 corresponds to a GRE score of 600: each falls one standard deviation above the mean of the distribution from which the score comes. Similarly, an IQ of 85 and a GRE score of 400 are each one standard deviation below the mean. An IQ of 130 corresponds to a GRE score of _23_; an IQ of 145 corresponds to a GRE score of _24_; an IQ of _25_ corresponds to a GRE score of 300; and an IQ of _26_ corresponds to a GRE score of 200. (Figure 6.12 on p. 105 illustrates these correspondences.)

23.
24.
25.
26.

There is a convenient way of saying where a raw scores falls within its distribution, a way that makes clear the sort of correspondences noted above. The raw score is converted to a _27_ score. The _28_ score states the position of the raw score in relation to the _29_ of the distribution from which the raw score comes, using the _30_ of the distribution as the unit for specifying the distance from the mean. Thus an IQ of 115 in the distribution of Wechsler IQs has a z score of $+1$, because it is 1 standard deviation (1 15-point unit) above the mean (which is 100). A GRE score of 600 within its own distribution also has a z score of $+1$, because it too lies one standard _31_ above the _32_, but here the standard deviation and the mean are those characterizing the distribution of GRE scores, namely 100 and 500, respectively. The z score for an IQ of 85 or a GRE score of 400 is _33_, and the z score for an IQ of 130 or a GRE score of 700 is _34_. The IQ with a z value of $+3$ is _35_, and the GRE score with a z value of -3 is _36_, as Figure 6.12 shows.

27.
28.
29.
30.

31.
32.

33.
34.
35.
36.

z scores can easily be computed in your head if the numbers involved are simple, like those in the examples above. If you need a formula to find a z score, in words it is z equals (raw score minus _37_) divided by _38_. In symbols, the formula is $z = $ _39_.

37.
38.

The two distributions in the examples above would both be close to normal in their shape, as Figure 6.12 shows, but _40_ scores are useful in describing raw scores in a distribution with any kind of shape.

39.
40.

Using z Scores to Solve Problems Involving the Normal Curve

In cases in which the normal curve is a good model for the _41_ of a distribution, z scores and the table in Appendix F of the text permit the solution of many kinds of problems. Sections 6.4 and 6.5 provide the details.

41.

Properties of z Scores

If all the raw scores in any given distribution are converted to z values, the mean of the z scores will always be _42_, and the standard deviation of the z scores will always be _43_. A distribution of z scores does not necessarily have a normal shape. It has whatever shape is characteristic of the set of raw scores from which the z scores were derived.

42.
43.

Comparing Scores from Different Distributions

To compare a raw score from one distribution with a raw score from a different distribution, it is helpful to convert each raw score to a _44_ score. Each _45_ score will, of course, show where the corresponding raw score lies in relation to the _46_ of its distribution, using as the unit of distance from the _47_ the standard _48_ of that distribution.

    Each z score will thus indicate where the corresponding raw score lies in relation to other scores in the distribution. But those other scores were earned by a specific group of people (they are called a _49_ group). And if the people who contributed scores to one distribution differ markedly from those who contributed scores to the other distribution, then a comparison of z scores from the two distributions may be misleading. z scores may properly be compared only when the _50_ groups used to generate the raw scores are comparable.

44.
45.
46.
47.
48.

49.

50.

Other Kinds of Standard Scores

There are three minor inconveniences associated with the use of z scores. First, both positive and _51_ values will occur in any given distribution. (The positive values will be for raw scores that lie _52: above or below?_ the _53_, and the negative values will be for raw scores that lie _54: above or below?_ the _55_. There must be at least one raw score above the mean and one below it for the reason explained in the section "Catch Your Own Errors" in Ch. 4 of this study guide.)

    Second, it is generally necessary to report z scores to one or two decimal places (as numbers like 2.1 or 2.13, for example) in order to be sufficiently precise in indicating the location of the corresponding raw score.

    And third, the public is unfamiliar with z scores, so it can be difficult to communicate to the public in terms of z scores.

51.
52.
53.
54.
55.

To circumvent these difficulties, statisticians have devised other kinds of scores like the *z*. These and the *z* variety are all called standard scores. In each case, the original raw scores are all converted to new values in a way that yields a distribution of new values with a specific mean and a specific standard deviation. For example, raw scores on the Graduate Record Exam are converted to a type of standard score in such a way that the distribution of standard scores for the population who take this test has a mean of __56__ and a standard deviation of __57__ .

56.

57.

Centile Scores* and the Normal Distribution

Standard scores, including *z* scores, are usually more meaningful than the original raw scores, because they indicate where a raw score lies in comparison to the scores earned by a known __58__ group. The centile ranks introduced in Ch. 3 function in the same way. For example, a person with a particular raw score on a certain test may have a centile rank of 63 in a distribution of the scores earned by some reference group; this centile score of 63 indicates that the person's performance exceeded that of __59__ % of the __60__ .

58.

59.

60.

Centile ranks are easy for the general public to comprehend, but they have a serious flaw. The difference between two centile ranks can give a misleading impression of the difference between the two corresponding raw scores. If the raw scores are distributed in a shape that is normal or at least roughly so, then a predictable kind of distortion occurs when raw scores are expressed as centile ranks. Look at Figure 6.12 on p. 105 for insight into this matter. Suppose we want to compare two GRE scores that are 100 points apart. If the scores are 500 and 600, the centile ranks are __61__ and __62__ , respectively, and these two centile ranks are 34 points apart. But if the GRE scores are 700 and 800, the centile ranks are __63__ and __64__ , which are not even two points apart. Comparing just the two centile ranks, then, can be misleading. In general, expressing scores in terms of __65__ suggests a relatively great difference in performance between two persons who are both near __66: the center or an extreme?__ of a distribution that is normal or approximately so, but it suggests a relatively small difference in performance between two persons who are both in __67: the center or an extreme?__ of such a distribution.

61.

62.

63.

64.

65.

66.

67.

---

*Section 6.8 of the text uses the term *centile score* to refer to what Ch. 3 called centile ranks.

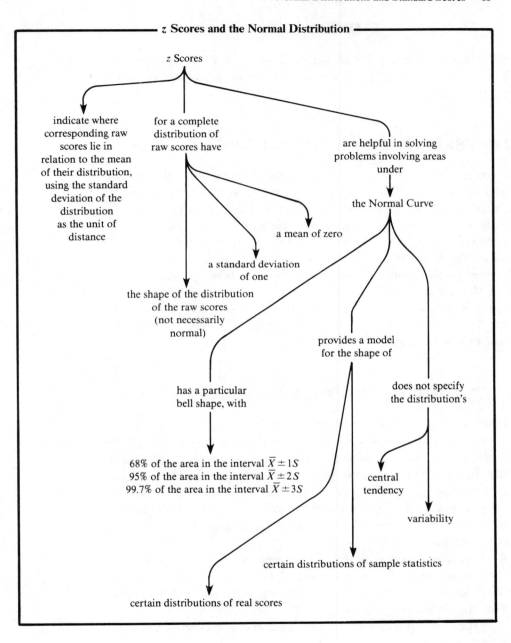

z Scores and the Normal Distribution

z Scores

indicate where corresponding raw scores lie in relation to the mean of their distribution, using the standard deviation of the distribution as the unit of distance

for a complete distribution of raw scores have

are helpful in solving problems involving areas under

the Normal Curve

a mean of zero

a standard deviation of one

the shape of the distribution of the raw scores (not necessarily normal)

provides a model for the shape of

has a particular bell shape, with

does not specify the distribution's

68% of the area in the interval $\overline{X} \pm 1S$
95% of the area in the interval $\overline{X} \pm 2S$
99.7% of the area in the interval $\overline{X} \pm 3S$

central tendency

variability

certain distributions of sample statistics

certain distributions of real scores

ADVICE

All of the exercises that your text offers are helpful if you want to develop expertise in statistics. But Problems 6 through 16 at the end of Ch. 6 in the text are especially important. Be sure you do them.

LABELING SKETCHES OF A NORMAL DISTRIBUTION

The text urges you to make a sketch of a normal distribution in working problems of the kind this chapter poses. Until you are *really* good at these problems, a sketch is absolutely essential, and it should be labeled carefully. *Above* the base line show *only* areas (as proportions or percentages). *Below* the base line, show *only* scores. Since you must work in terms of *z* scores, below the base line there should always be one or more scores of this kind. If the problem also involves raw scores, add them carefully, *below* the line and neatly aligned with the corresponding *z* scores.

The point of this advice is to keep you from confusing a proportion or a percentage of area with a score. If you follow this principle, areas will always be above the base line, and scores always below. The figures in the text provide good models illustrating these principles.

WHERE TO GO FOR HELP

As you have surely observed by now, the study of statistics is like climbing a ladder: you progress up a series of steps, and you can get to a higher step only through the lower ones. Thus if you're unsure of some previous matter, you may not be able to learn what's new. In this chapter, for example, if you lack a clear understanding of the mean and the standard deviation, you'll fail to understand *z* scores.

What can you do if you need to review a topic in order to progress? Look back in the text, obviously, but looking ahead may also help, if you look all the way ahead to the appendices and the index.

Appendix A offers you a review of basic mathematics; Appendix B lists the symbols that appear in the text; and Appendix C is a collection of important formulas. All three can be very helpful. Look them over now, so you'll remember they're there if you need their help later.

Remember also the index; it can point you quickly back to the information you need on a specific topic.

And there's always this *Guide*, of course. You may benefit from reviewing whatever this book has to say on a matter that's puzzling you.

EXERCISES

To practice locating scores relative to the mean of their distribution, using the standard deviation as the unit of distance from the mean, fill in the mssing values in the table below. Remember that the four scores on a line would be truly equivalent only if the distributions from which they come have similar shapes. You should be able to do most of the calculations for the first four columns in your head.

TABLE OF EQUIVALENT SCORES

z Score	Score where $\bar{X}=100, S=15$	Score where $\bar{X}=500, S=100$	Score where $\bar{X}=50, S=10$	Centile Rank if Shape is Normal
+3.00	145			
				97.72
		650		
				50.00
−1.00				
		250		
			20	

SPECIAL PROPERTIES OF A DISTRIBUTION OF z SCORES

Here's a collection of scores listed in order from high down to low. (This is not a frequency distribution.) Take them to be a sample called X. Compute the mean, find the deviation scores, and then calculate the standard deviation from the squares of the deviation scores. These are the sorts of exercises the preceding chapter of this *Guide* offered you.

What's new here is the opportunity to see how the mean of a distribution of z scores works out to be zero and how the standard deviation works out to be one (when the arithmetic is correct). To enjoy this elegant simplicity, find the z score corresponding to each raw score. Check to be sure that the mean of the z scores, $\bar{z}$ ("zee bar"), is zero. Then compute the standard deviation of the z scores, S_z ("ess sub zee"), using a procedure like that by which you found the standard deviation of the raw scores. This requires treating the z scores as though they were raw scores and finding the deviation scores for them, using their own mean, zero, as a reference point.

All the figures in the table work out to be simple (but not necessarily whole) numbers—if you do no rounding. A completed version of the table appears at the back of this book, in case you have trouble with the arithmetic.

In doing these computations, as in doing all computations in statistical work, you should *think what the numbers mean*. In finding the mean of a column of numbers, for example, remember that what you get when you divide the total by the n is supposed to be an average of those numbers. It's supposed to indicate about how big they run, and it must fall somewhere between the lowest score and the highest. Staying aware of what you're doing in this way will permit you to catch some errors that you might otherwise not notice. The preceding chapter of this *Guide* offered a list of thoughts worth thinking in doing statistical computations; you may wish to review it at this time.

Raw Score	Deviation Score	Squared Deviation Score	z Score	Deviation Score for z Score	Squared Deviation for z Score
X	$x = (X - \bar{X})$	x^2	$z = x/S$	$(z - \bar{z})$	$(z - \bar{z})^2$
13					
13					
9					
6					
4					
3					

$\Sigma X =$ $\Sigma x =$ $\Sigma x^2 =$ $\Sigma z =$ $\Sigma(z - \bar{z})$ $\Sigma(z - \bar{z})^2 =$

$n =$ $S^2 = \Sigma x^2/n$ $n =$ $=$ $S_z^2 = \Sigma(z - \bar{z})^2/n$

$\bar{X} = \Sigma X/n$ $= \quad /$ $\bar{z} = \Sigma z/n$ $= \quad /$

$= \quad /$ $=$ $= \quad /$ $=$

$=$ $S = \sqrt{}$ $=$ $S_z = \sqrt{}$

$=$ $=$

THE NORMAL DISTRIBUTION AND *z* SCORES IN USE

A psychologist who has conducted important research on mental retardation recently discussed the issue of how to define retardation. Writing in the journal *Science*, Edward Zigler of Yale University notes that in 1959, the American Association on Mental Deficiency, the leading professional organization concerned with the retarded in America, defined the retarded as those with an IQ score one or more standard deviations below the mean.

_____1. What IQ score is one standard deviation below the mean? (The IQ tests for children have a mean and standard deviation that are the same as those for the WAIS or very close to it.)

_____2. What percentage of the population was defined as retarded by this criterion?

But many professionals now accept an IQ of 70 as the dividing line, Zigler says.

_____3. How many standard deviations below the mean does an IQ of 70 lie?

_____4. By this criterion, what percentage of the population is retarded?

Zigler argues that retardation is a "two-tiered phenomenon" in which the "mildly retarded," those with IQ between 50 and 70, must be distinguished from the "more severely retarded," those with IQ less than 50. "Statements or views concerning one tier are pretty much inapplicable to the other," Zigler believes.

_____5. How many standard deviations below the mean does an IQ of 50 lie?

_____6. What percentage of the population is "mildly retarded" according to Zigler's definition?

_____7. What percentage of the population is "severely retarded"?

Zigler suggests that people with IQ between 50 and 70 who are currently called retarded merely represent "the lower part of the normal population." In this conception, these people are like those whom we regard as short. Most of the people in each classification came to be what they are through the usual genetic and environmental processes, which produce a normal distribution of IQ scores in the one case and a normal distribution of heights in the other, both of them just naturally including some scores well below the mean. The truly retarded, in Zigler's view, are like the dwarfs and midgets who fall far below the mean in height as a result of processes that are clearly abnormal.

Unfortunately, he notes, there is no emotionally neutral word comparable to "short" to describe the lower end of the naturally occurring distribution of IQ scores. The word "retarded" in pejorative, and to apply it to all people with an IQ between 50 and 70 is unfair and misleading, Zigler believes, just as it is unfair and misleading to refer to all adults who are, say, five-two or less as dwarfs or midgets.

IQ is not the only criterion that is used in diagnosing retardation. The person's social competence and the age at which the abnormalities began are also taken into consideration by many professionals. Social competence, or the ability to meet the demands of everyday living, is not adequately defined by an IQ score, Zigler says, and the exact relation between intelligence and social competence is unclear. There is great need for a measure of social competence that can be used throughout the lifespan as IQ tests can.

Reference: E. Zigler, Review of *The Mentally Retarded and Society*, *Science*, 1977, *196*, 1192–1194.

SELF-TEST

The normal curve is a model for which characteristic(s) of a distribution of numbers?   A) Its central tendency.   B) Its variability.   C) Its shape.   D) All of these.

1.   A   B   C   D

The normal curve is useful in ...   A) descriptive work.   B) inferential work.   C) neither of these.   D) both of these.

2.   A   B   C   D

About two-thirds of the area under the normal curve lies in the interval ...   A) $\overline{X} \pm 1S$.   B) $\overline{X} \pm 2S$.   C) $\overline{X} \pm 3S$.   D) $\overline{X} \pm 4S$.

3.   A   B   C   D

In a distribution of numbers whose shape is normal ...   A) the mean is always zero.   B) the standard deviation is always one.   C) the centile rank of the mean is never 50.   D) the mean, median, and mode always take the same value.

4.   A   B   C   D

In a distribution of 80 scores, the mean is 25. What is the z score that corresponds to a raw score of 30?   A) $+5$ B) -5 C) $+0.06$ D) It cannot be computed without more information.

5.   A   B   C   D

A raw score below the mean of its distribution has a z score that is ...   A) negative.   B) positive.   C) below -1.   D) larger than the raw score's centile rank.

6.   A   B   C   D

If a distribution of reaction times is positively skewed, what will be the shape of the distribution of z scores corresponding to the raw scores? A) Normal. B) Bell-shaped, but not necessarily normal. C) Negatively skewed. D) Positively skewed.

7.   A   B   C   D

Which type of score does *not* indicate where a person stands in comparison to others? A) A centile rank. B) A raw score. C) A z score. D) A T score.

8.   A   B   C   D

If a distribution of heights is normal in shape, what is the centile rank of the mean? A) Below 50. B) 50. C) Above 50. D) It cannot be computed without more information.

9.   A   B   C   D

Wechsler IQ scores have a mean of 100 and a standard deviation of 15. GRE scores have a mean of 500 and a standard deviation of 100. What GRE score corresponds to an IQ of 70? A) -2. B) 300. C) 470. D) 700.

10.   A   B   C   D

CHAPTER 7

CORRELATION

OBJECTIVES

Thus far in your study of statistics, there has been only one score for a given subject, a score indicating the value of just one variable (the subject's height, for example). When two scores are available for each subject, one score for each of two variables (the subject's height *and* the subject's weight, say), it is possible to determine the correlation between the two variables for the subjects on hand. Correlation is the topic of Ch. 7. In studying this chapter, work to understand the following symbols, terms, concepts, principles, and techniques:

Bivariate Distribution
Scatter Diagram
Linear Relationship
Pearson r
    Magnitude of r as an Index of Strength of Association   •   Sign of r as an Index of Direction of Correlation   •   Cross Products   •   Effect on r of Location of Points in Quadrants in Scatter Diagram
Correlation is not Causation
Factors Affecting r
    Linearity of Regression vs. Curvilinearity   •   Restriction of Range   •   Units of Measurement   •   Random Sampling Variation
There is No Such Thing as *the* Correlation Coefficient for Two Variables
    Effects of: Restriction of Range   •   Random Sampling Variation   •   Circumstances of Study
Measures of Strength of Association
    Size of r Relative to Zero and One   •   Scatter Diagram   •   Proportion of Variance Explained: r^2

Also in studying this chapter, work to learn these skills:

Construction of a Scatter Diagram
Calculation of Pearson r
    From Deviation Scores   •   From Raw Scores
Calculation of Proportion of Variance of One Variable Associated with the Other

REVIEW

Bivariate Distributions and Scatter Diagrams

A problem in correlation begins with a collection of __1__ scores. For each subject, there is a first score that indicates his or her standing on a first variable, and paired with this score is a second that indicates the subject's standing on a second variable. The first variable is usually called __2__, and the second is usually called __3__. Look at the first three columns of the table on p. 117 for an example of this kind of data. (But note that here the columns of raw scores are labeled "1st Quiz" and "2nd Quiz" rather than X and Y.) Scores that are paired like these constitute a __4__ distribution.

The scores of a bivariate __5__ can be displayed in a graph called a __6__. Look at Figure 7.2 on p. 118 for the graph of the __7__ distribution on p. 117. In such a graph, each dot represents an individual subject (case). The dot indicates simultaneously how the subject scored on the first variable, X, *and* how she or he scored on the second variable, __8__.

Scatter __9__ for other bivariate __10__ appear in Figure 7.1 on p. 116, and they illustrate various possible patterns.

1.

2.

3.

4.

5.

6.

7.

8.

9.

10.

Correlation: A Matter of Direction

There is *no* correlation between two variables if high scores on one variable tend to be associated with *both* high and low scores on the other variable, and if low scores on the one variable also tend to be associated with both high and low scores on the other. Graph 4 of Figure 7.1 on p. 116 illustrates a case of a correlation very close to zero.*

If there *is* a correlation between two variables, it will be either positive or __11__. In the case of a *positive* correlation, high scores on one variable tend to be associated *only* with __12: high or low?__ scores on the other variable, while low scores on the one variable tend to be associated only with __13: high or low?__ scores on the other. There is a positive correlation between height and weight among human beings, in that people with a high score on height (tall people) tend to have a __14: high or low?__ score on weight (tend to be heavy), while people with a low score on height (short people) tend to have a __15: high or low?__ score on weight (tend to be light). Positive correlations are illustrated in Graphs 1 - 3 of Figure

11.

12.

13.

14.

15.

*In Figure 7.1, take the graphs in the top row to be Nos. 1 and 2, the graphs in the middle row to be Nos. 3 and 4, and those in the bottom row to be 5 and 6.

7.1. In these diagrams, the data points fall in a swarm running from lower left to _16_ .

16.

In the case of a *negative* correlation, high scores on one variable tend to be associated only with _17: high or low?_ scores on the other variable, and low scores on the one tend to be associated only with _18: high or low?_ scores on the other. There is a negative correlation between height and normal nightly sleeping time among human beings, in that people with a high score on height (adults) tend to have a _19: high or low?_ score on nightly sleeping time (tend to sleep relatively few hours), while people with a low score on height (children) tend to have a _20: high or low?_ score on nightly sleeping time (tend to sleep relatively many hours). Negative correlations are illustrated in Graphs 5 and 6 of Figure 7.1 In these diagrams, the data points fall in a swarm running from upper left to _21_ .

17.

18.

19.

20.

21.

Correlation: A Matter of Degree

The distinction between a positive and a negative correlation is a distinction in the *direction* of the correlation. Correlations also vary in *degree*. In the case of a *perfect* correlation, all data points in the scatter _22_ fall on a _23_ line. (If the *direction* is positive, the line slopes from lower left to _24_ , as in Graph 1 of Figure 7.1; if the *direction* is negative, the line slopes the other way, as in Graph 6.) In the case of less than perfect correlations, the data points swarm more or less closely about a straight line; the farther away the data points are, the _25_ the degree of the correlation (whether the *direction* is positive or negative).

22.

23.

24.

25.

Pearson's *r* as a Measure of Correlation

The direction and the degree of correlation between two variables are indicated by a number called the Pearson correlation _26_ . The coefficient is symbolized _27_ for a bivariate _28_ that is a sample.

The sign of the coefficient may be positive or _29_ . A positive value of *r* indicates a positive correlation, which, as noted above, occurs when there is a tendency for high values of one variable (X) to be associated with _30: high or low?_ values of the other variable (Y), and low values of the one to be associated with _31: high or low?_ values of the other. A negative value of *r* indicates a _32_ correlation, which occurs when high values of X are associated with _33_ values of Y, and vice versa. The sign of the coefficient thus indicates the _34: direction or degree?_ of the association.

26.

27.

28.

29.

30.

31.

32.

33.

34.

When no relationship between the two variables exists, the value of r is __35__. When a perfect relationship exists, the value of r is __36__ (plus or minus, depending on the direction of the correlation). An intermediate degree of relationship is represented by an intermediate value of __37__.

When r is $+1.00$ or -1.00, every point of the scatter diagram lies exactly on a __38__ line. This means that if we know the value of X, we can predict the value of Y with no error. When r is between 1.00 (plus or minus) and zero, the points on the scatter diagram do not lie exactly on a straight line. In such cases, if we know the value of X, we can predict the value of Y only with some error. The next chapter tells you in detail just how r figures into attempts to predict Y from X.

The Pearson r is defined in the text in terms of deviation scores. The formula is $r_{XY} = $ __39__. In this formula, $x = (X - \overline{X})$, and $y = $ __40__; these quantities are the deviation scores. Σxy is the sum of the products of the paired __41__ scores; it is sometimes called the sum of the cross products of the deviation scores. The symbol n stands for the number of __42__ of scores. S_X is the __43__ of the scores called X, and __44__ is the standard deviation of the scores called Y.

To understand how this formula works, look at Figure 7.2 on p. 118. This __45__ diagram is divided into four quadrants by two lines, one located at the __46__ of X and one at the __47__ of Y. Points that lie to the right of the vertical line are therefore characterized by __48: positive or negative?__ values of x and those to the left by __49: positive or negative?__ values of x. Those points lying __50: above or below?__ the horizontal line are characterized by positive values of y, and those __51: above or below?__ by negative values of y. For any point, the xy product may be positive or negative, depending on the sign of x and the sign of y. The xy products will be positive for points falling in quadrants __52__ and will be negative for points falling in quadrants __53__.

If you inspect the deviation-score formula for r, you can see that the __54__ of the fraction, Σxy, determines whether the value of r will be negative, zero, or positive. When the data points are concentrated primarily in quadrants __55__, the positive xy products will exceed the negative xy products from quadrants __56__, and Σxy will be positive; therefore r will be positive. On the other hand, when the points are concentrated primarily in quadrants __57__, the negative xy products will exceed the positive xy products from quadrants __58__, and __59__ will be negative; therefore r will be __60__.

As for the *magnitude* of r, that is determined by the extent to which the points are concentrated in just two of the quadrants (either in I and III or in II and IV). The more the points (cases) are concentrated in two of the

35.
36.
37.

38.

39.
40.
41.
42.
43.
44.

45.
46.
47.
48.
49.
50.
51.

52.
53.
54.

55.
56.

57.
58.
59.
60.

quadrants, the more the xy products of one sign will exceed those of the other sign, and the larger the magnitude of Σxy will be (either positive or negative). The larger the magnitude of Σxy, the larger __61__ will be (either positive or negative).

61.

Correlation and Causation

It is important not to confuse correlation with causatior. Even a substantial correlation coefficient between two variables does not necessarily mean that a __62__ relationship exists between the two. When we find two variables, X and Y, to vary together, there are several possible explanations for the covariation. Variation in X could be responsible for variation in __63__; variation in Y could be responsible for variation in __64__; or there could be a third factor (or complex of factors) that affects both X and Y. Merely knowing there is a correlation between the two variables does not provide an explanation for the correlation. Thus if we are to infer causation, it must be on logical grounds over and above the statistical demonstration of __65__.

62.

63.
64.

65.

Factors Influencing the Correlation Coefficient

1. *Linearity of regression:* Ch. 7 is concerned with cases in which the points of a scatter diagram are well fitted by a line that is straight. In such cases, the data (the pairs of scores, the bivariate distribution) are said to possess the property of linearity of __66__. The Pearson r is a good index of the direction and degree of correlation only in cases of this kind. When the points of the scatter diagram closely hug a *curved* line, the relationship between the two variables is strong, but it is not linear; rather it is __67__. To the extent that a relationship is curvilinear, r will __68: over- or underestimate?__ the true degree of association between the two variables.

66.

67.
68.

2. *Restriction of range:* As noted above, the magnitude of the correlation __69__ depends on the nature of the relationship between the two variables—specifically, on the extent to which the points in the scatter diagram hug a straight line. But the magnitude also depends on the degree of variability among the X values or among the Y values (or both). Other things being equal, the greater the restriction of range (variability) in either X or Y or both, the __70: higher or lower?__ the correlation coefficient will be.

69.

70.

3. *Units of measurement:* The Pearson __71__ is completely unaffected by transformations of the scores on either variable from one set of units to

71.

another, so long as the transformation is done by adding a constant to each score, subtracting a constant from each score, or multiplying or dividing each score by a constant. Why is this so? Because the Pearson r reflects the degree to which *relative* positions on X match up with 72 positions on 73 . And the type of transformation that involves applying a constant to each score on X or each score on Y (or both) does not change the relative positions of the cases.

4. *Sampling variation:* Correlations are affected by the 74 factors involved in the selection of a sample. The value of r for a sample will therefore almost always differ to some extent from the true correlation (the correlation between the two variables for the whole population from which the sample came). The larger the sample, the 75: more or less? likely it is that the value of r departs strongly from the true value.

72.

73.

74.

75.

Is There Such a Thing as *the* Correlation Coefficient?

There is no such thing as *the* correlation coefficient for two variables. Two factors that influence the value we obtain were noted just above: the extent to which the range (variability) in X or Y or both is 76 , and the extent to which 77 produced a sample of paired scores with characteristics different from those of the population of paired scores. Furthermore, the degree of association between two variables also depends on: 1) the specific way in which each variable is measured; 2) the kinds of 78 studied in the investigation; and 3) the particular circumstances under which the variables are at work. Consequently, it is of utmost importance that a correlation coefficient be interpreted in the light of the particular conditions under which it was obtained.

76.

77.

78.

Strength of Association

As already noted, the magnitude of the Pearson r indicates the degree of the linear (straight-line) association between the two variables. But exactly how is the magnitude of the r value to be interpreted? There are several possibilities.

1. *Relative to the theoretical limits of zero and one:* Values of r close to 79 indicate very little linear association, while values close to 80 (plus or minus) indicate a high degree of linear association. That is, values of r close to 81 indicate that the data points in the scatter diagram fail to hug a straight line closely, while values of r close to 82 indicate that the data points do hug such a line. But the degree of linear (83 -line)

79.

80.

81.

82.

83.

association is not necessarily interpretable in direct proportion to the magnitude of the coefficient. For example, a correlation of .50 indicates ___84: more or less?___ than half the strength of association shown by a correlation of 1.00.

84.

2. *In terms of the pattern shown by the* __85__ : The scatter diagram provides a literal picture of the association.

85.

3. *In terms of the proportion of variance explained:* In this way of interpreting the magnitude of the correlation coefficient, we must think of variation in terms of the __86__ rather than in terms of the standard __87__ . The total variance in either variable can be broken down into component parts: the part associated with variation in the other variable, and the remainder. The proportion of the total variation that is associated with variation in the other variable is given by the __88__ of the correlation coefficient.

86.
87.

88.

ESTIMATING CORRELATIONS

For each case below, use your common sense and your new understanding of correlation to determine whether the correlation between the two sets of scores is zero, positive, or negative; and if positive or negative, whether the degree of correlation is low, medium, high, or perfect.

_____1. Among the world's population of human beings with both legs intact, what is the correlation between the length of a person's left leg and the length of his or her right leg?

_____2. Among grade-school children, what is the correlation between the length of a child's nose and the number of words in his or her spoken vocabulary? Remember that the children come from grades one through six.

_____3. Among teenagers, what is the correlation between the number of freckles on a person's face and the person's IQ?

_____4. Among married couples in North America, what is the correlation between the age of the husband and the age of the wife?

_____5. *Playboy* magazine, following a suggestion offered by Plato 2300 years ago, once proposed that men should marry about

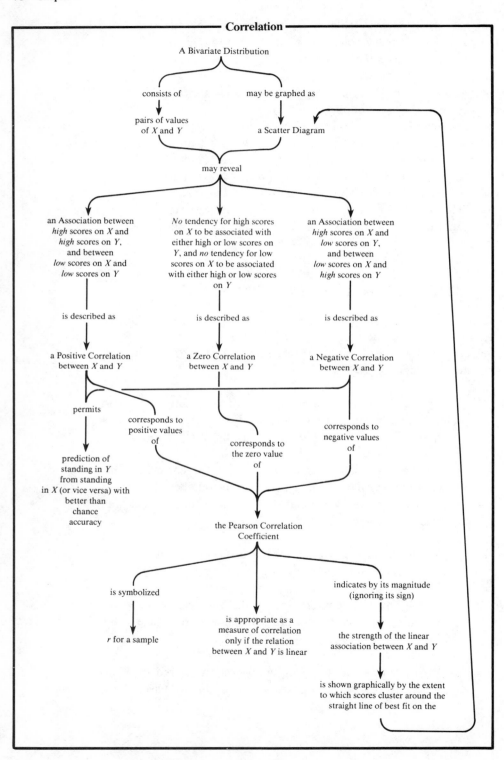

Correlation

A Bivariate Distribution

consists of

may be graphed as

pairs of values
of X and Y

a Scatter Diagram

may reveal

an Association between
high scores on X and
high scores on Y,
and between
low scores on X and
low scores on Y

No tendency for high scores
on X to be associated with
either high or low scores on
Y, and *no* tendency for low
scores on X to be associated
with either high or low scores
on Y

an Association between
high scores on X and
low scores on Y,
and between
low scores on X and
high scores on Y

is described as

is described as

is described as

a Positive Correlation
between X and Y

a Zero Correlation
between X and Y

a Negative Correlation
between X and Y

permits

corresponds to
positive values
of

corresponds to
the zero value
of

corresponds to
negative values
of

prediction of
standing in Y
from standing
in X (or vice versa) with
better than
chance
accuracy

the Pearson Correlation
Coefficient

is symbolized

indicates by its magnitude
(ignoring its sign)

is appropriate as a
measure of correlation
only if the relation
between X and Y is linear

r for a sample

the strength of the linear
association between X and Y

is shown graphically by the extent
to which scores cluster around the
straight line of best fit on the

the age of 30 and choose a bride about the age of 20. Suppose that North America follows this advice for 100 years. After the 100 years, married couples are surveyed; some are relatively young newly-weds, and some are just celebrating their golden anniversary. Among these couples, what would be the correlation between the age of the husband and the age of the wife?

_____6. What is the correlation between the height of the New York Yankees baseball players and the New York Jets football players?

SELF-TEST

A bivariate distribution consists of ...   A) two-digit scores.   B) pairs of scores.   C) scores for two subjects.   D) a distribution with two modes.

1.   A   B   C   D

A scatter diagram is ...   A) a table.   B) a graph.   C) an inferential technique.   D) a distribution with high variability.

2.   A   B   C   D

If the points in a scatter diagram slope from lower left to upper right, ... A) the correlation is positive.   B) the correlation is zero.   C) the correlation is negative.   D) it is impossible to compute the correlation coefficient.

3.   A   B   C   D

The magnitude of the correlation coefficient (ignoring its sign) indicates ...   A) the strength of the association between the two variables. B) the extent to which the points in the scatter diagram hug a straight line.   C) both of these.   D) neither of these.

4.   A   B   C   D

When there is no relationship between two variables, ...   A) r is positive. B) r is zero.   C) r is negative.   D) it is impossible to compute r.

5.   A   B   C   D

We can properly conclude that Variable X causes Variable Y ...   A) if research shows that X is positively correlated with Y.   B) if research shows that X is negatively correlated with Y.   C) in both of these cases. D) in neither of these cases.

6.   A   B   C   D

Suppose the correlation between IQ and score on a test of creativity is .52 for a large sample of adults from all walks of life. What would be the probable value of r for just those people whose IQ is more than two standard deviations above the mean IQ?   A) Lower than .52.   B) About .52, but not necessarily exactly .52.   C) Exactly .52.   D) Greater than .52.

7.   A   B   C   D

Suppose a statistics instructor gives his or her students a test of computational accuracy, scores it as the number of items correct, and correlates these scores with the number of items the students correctly answered on the first exam in the statistics course. Suppose further that the Pearson r turns out to be .38. If the instructor rescores the test of computational accuracy as the proportion correct and then recalculates the value of r, the instructor will find that ... A) r is less than .38. B) r is about .38, but not necessarily exactly .38. C) r is exactly .38. D) r is greater than .38.

8.   A   B   C   D

Suppose a scatter plot shows that the points lie in a pattern with the shape of an upside-down **U**. If we calculate r for these data, we will find that it ... A) is positive. B) is negative. C) overestimates the closeness of the association between the two variables. D) underestimates the closeness of the association between the two variables.

9.   A   B   C   D

Which value for r most nearly indicates a correlation whose degree is twice as strong as the degree of correlation indicated by a value of .5? A) $-.7$. B) .25. C) 1.00 D) It cannot be determined without more information.

10.   A   B   C   D

CHAPTER 8

REGRESSION AND PREDICTION

OBJECTIVES

When two variables are correlated, it is possible to predict, with better than chance accuracy, standing in one from knowledge of standing in the other. Ch. 8 explains how to make such predictions. In studying the chapter, work to understand the following symbols, terms, concepts, principles, and techniques:

Correlation as the Basis for Prediction
Prediction from a Graph
Y' • Error • The Regression Line • The Least Squares Criterion • Y' as an Estimated Mean • The Other Line of Best Fit
Prediction from an Equation
The Raw-Score Formula
The Standard-Score Formula
$z_{Y'}$ • z_X
Prediction when $r = +1$
Regression toward the Mean
Prediction when $r = 0$
Prediction when $X = \bar{X}$
The Standard Error of Prediction: S_{YX}
As a Measure of the Variability of Actual Y Values about Y'
In Comparison to S_Y
Formula Based on r
Limits of the 95% Confidence Interval for the True Value of Y
Homoscedasticity
The Relation between r and Predictive Error

Also in studying this chapter, work to learn these skills:

Prediction of Y from a Graph Showing the Regression Line
Construction of the Raw-Score Formula for Y'
Prediction of Y from the Raw-Score Formula

81

Construction of the z-Score Formula for $z_{Y'}$
   Prediction of $z_{Y'}$ from the z-Score Formula
Calculation of S_{YX}
   From Y and Y' •   From S_Y and r
Construction of a 95% Confidence Interval for the True Value of Y

REVIEW

The Regression Line

Look at Figure 8.1 on p. 136 to see how predictions are made in terms of a graph. This kind of graph is called a __1__. Each dot on the graph represents a subject. The dot for a given subject tells us the subject's score on two variables, one called X and the other called __2__. The position of the dot over the horizontal axis indicates the score on __3: X or Y?__, and the position of the dot along the __4: horizontal or vertical?__ axis indicates the score on Y.

   When the swarm of data points is reasonably described by a straight line, as in this figure, the data are said to show __5__ of regression. (See Section 7.7 on pp. 121 - 122 for a review of this matter). In such a case it is possible to find the straight line that best fits the data points and use it to predict values of one variable from values of the other. By convention, the predictor variable is the one called __6: X or Y?__ that is represented on the __7: horizontal or vertical?__ axis, and the variable to be predicted is the one called __8: X or Y?__ that is represented on the __9: horizontal or vertical?__ axis. The prediction is made by starting with the value of X along the horizontal axis, going straight up to the line, and then going straight over to the vertical axis. The predicted score on the vertical axis is symbolized __10__ ("wye prime"). The line is called the __11__ line.

1.

2.

3.
4.

5.

6.
7.
8.
9.

10.
11.

Errors in Prediction and the Location of the Regression Line

If the correlation coefficient is $+1$ or __12__, each point on the scatter diagram will fall __13: close to or exactly on?__ the regression line, and prediction will be errorless. With a correlation whose degree is less than perfect, there will necessarily be some error in prediction. (See "Correlation: A Matter of Degree" in Ch. 7 of this *Guide* if you wish to review the concept of degree of correlation and the way in which r reflects the degree.) The error for a given case is the discrepancy between the actual score on Y and the predicted score. In symbols, error = __14__. The

12.
13.

14.

regression line is located in such a way that the __15__ of the squared   15.
predictive errors for all cases on hand is as small as possible. This
criterion for locating the best fitting line is known as the least __16__   16.
criterion.

The Regression Line as a Running Mean

If linearity of regression holds, the regression line is a kind of "running
mean." The cases with a given score on X will have a variety of scores on
Y, and the predicted value of Y obtained from the regression line for
these cases is an estimate of the __17__ of their Y values. See Figure 8.3 on   17.
p. 138 for an illustration.

The Regression Equation: Raw Score Form

Every straight line on a graph has an equation. The equation of the
regression line is known as the __18__. In terms of raw scores, the equation   18.
is complicated:

$$Y' = r(S_Y/S_X)X - r(S_Y/S_X)\overline{X} + \overline{Y}$$

In this equation, Y' is the __19: predicted or observed?__ raw score on Y, r is   19.
the coefficient of correlation between X and Y, S_Y is the __20__ of the Y 20.
values, __21__ is the standard deviation of the X values, X is the raw score   21.
from which prediction is to be made, $\overline{X}$ is the __22__ of the X values, and   22.
__23__ is the mean of the Y values.   23.

To use this equation for a given set of X and Y scores, we calculate $\overline{X}$
and $\overline{Y}$, S_X and S_Y, and r, and we insert their values into the equation,
leaving X unspecified. We then "boil down" the equation (as illustrated
at the top of p. 140), and we end up with a simple expression of the form
$Y' = bX + a$, where b and a are certain numbers.* We can then take any
value of X from which we wish to make a prediction and use the
simplified form of the equation to generate the value of __24__. This   24.
numerical procedure will be equivalent to the graphical one involving the
regression line on the scatter diagram.

If we ask the regression equation to predict the value of Y for a case
whose value of X is at the mean of X, the prediction will always be the
__25__ of Y. In graphical terms, the regression line always goes through the   25.
point at the intersection of the two means, which is represented by the
symbols $(\overline{X}, \overline{Y})$.

*In the terms usually applied to "linear equations" like this one, b is the slope of the line
and a is the Y-intercept.

The Regression Equation: Standard Score Form

The regression equation can also be stated in terms of the standard scores called z. (See Ch. 6 for a review of z scores.) In these terms, the regression equation is very simple:

$$z_{Y'} = 26$$

26.

In this equation, $z_{Y'}$ ("zee sub wye-prime") is the predicted z score on Y, r is the coefficient of __27__ between X and Y, and z_X ("zee sub eks") is the z score on X from which the prediction is made.

27.

This form of the __28__ equation illustrates some important aspects of prediction. First, when $r = +1$, the predicted standard score on Y is __29: identical to or less than?__ the standard score on X from which prediction is made. The top panel of Figure 8.4 on p. 142 illustrates this state of affairs.

28.

29.

Second, when the value of r is less than $+1$ (that is, when the correlation is less than perfect), a phenomenon called regression toward the mean occurs. We start a prediction with a value of X that lies a certain number of standard deviations away from the mean of X. (The z score, z_X, indicates what this number is. For example, a case that is two standard deviations below the mean of X has a z_X value of __30__. The "$-$" indicates that the case is below the mean, and the "2" indicates that it is two __31__ away.) We now predict that this case has a standing in Y that is a certain number of standard deviations away from the mean of Y; the predicted z score, $z_{Y'}$, tells us what this number is. And the distance between the predicted score and the mean of Y will be less than the distance between the X score and the mean of X (when the standard deviations are used as the units in which distance is measured). For example, if $r = .5$, what is $z_{Y'}$ for a z_X of -2, which indicates an X score two standard deviations below the mean? The equation says that $z_{Y'} = rz_X = (.5)(-2) = -1$, and a score whose z value is -1 is only __32__ standard deviation below the mean. The middle panel of the figure on p. 142 illustrates this phenomenon.

30.

31.

32.

To understand regression toward the __33__, it is important to remember that the Y' value for a given value of X is an estimate of the mean of the Y scores for all cases with the given X value. The prediction is thus going to be correct "on the average" (in general) but not in every case. So, because the correlation between the height of a parent and the height of a grown child is less than perfect, we can correctly predict that a relatively tall parent will *in general* have a child who, when grown, is also relatively tall, but who is not as tall relative to other grown children as the parent was relative to other parents. And because the correlation between

33.

mathematical aptitude and statistics grades is less than perfect, we can correctly predict that statistics students with the highest math aptitude scores will earn relatively high grades, though *in general* not the highest. But there can still be individual exceptions. A tall parent can have a child who when grown is even taller than the parent. A student with a high score on a mathematical aptitude test can do even better, relatively speaking, in a statistics course than she or he did on the aptitude test.

Third, when $r = 0$, no matter what X score we predict from, the predicted z score on Y will always be __34__, which indicates a raw score that is at the __35__ of Y. Thus if X and Y are uncorrelated, the best prediction of standing on Y for any value of X is the __36__ of Y.

34.
35.
36.

Fourth, for a case that is at the mean of X, with a z_X value that is accordingly zero, no matter what the correlation is, the regression equation will always predict a z score of __37__ on Y, indicating a prediction that this case is also at the mean on Y.

37.

Measuring the Margin of Predictive Error

When the correlation between X and Y is less than perfect, as is invariably the case in the real world, predictions from the __38__ equation will vary somewhat from the correct values. A measure of the variability is thus useful. Such a measure is the standard __39__ of prediction, symbolized __40__ ("ess sub wye eks"). The higher the value of S_{YX} for a given set of observations, the __41: larger or smaller?__ the errors in prediction.

38.
39.
40.
41.

S_{YX} is a kind of standard deviation, and this is appropriate, because like the standard deviation, it is a measure of __42__. But where the standard deviation measures variability about a mean such as $\overline{Y}$, the standard error of prediction measures variability about a predicted score Y'. The formula for the standard deviation of the actual Y values, symbolized S_Y, is __43__. The formula for the standard error of prediction, S_{YX}, is the same except that __44__ replaces __45__.

42.

43.
44.
45.

The formula that defines the standard error of __46__ is awkward to use for actually computing S_{YX} for a given set of data. Fortunately, mathematicians have discovered that an elegantly simple formula involving the correlation coefficient r will produce the same results. The formula is:

46.

$$S_{YX} = \text{\_\_47\_\_}$$

47.

If actual scores on the variable Y are normally distributed around the predicted values symbolized Y', then we can use what we know about the normal distribution to make some inferences about the errors of prediction. (See Ch. 6 for a review of the normal distribution.) We know, for example, that in any distribution whose shape is the one called normal, the middle 95% of the cases fall within the interval from 1.96 standard deviations below the __48__ to __49__ standard deviations above the __50__. In applying this knowledge to errors of prediction, we take the standard deviation to be S_{YX}, and we take the mean to be Y'. We therefore expect the middle 95% of the cases with a particular score on X to obtain Y scores within the limits $Y' - 1.96$ standard errors of prediction and $Y' + 1.96$ standard errors of prediction. In symbols, the limits for the middle 95% of the Y scores are $Y' \pm \underline{51}$.

48.
49.
50.

51.

Because the regression line obtained from a sample usually differs from the "true" prediction line for the entire __52__ from which the sample came, and because the difference is likely to be relatively large for a __53: large or small?__ sample, it is best to count on the procedures described here only when sample size is at least __54__.

52.
53.
54.

The Relation between *r* and Predictive Error

As the correlation between two variables becomes stronger (whether it is positive or negative), the value of r (ignoring its sign) becomes __55: larger or smaller?__, and the errors of prediction become __56: larger or smaller?__. Using the standard error of prediction, symbolized __57__, as a measure of the errors of prediction, we can be precise about exactly how an increase in the value of r reduces the predictive error.

55.
56.
57.

Look at the table on p. 147. At the bottom of the table, we see that when r is zero, there is __58__ % reduction in predictive error; that is, this is the worst possible case for making predictions. With a correlation of zero, the best we can do, no matter what score on X we are predicting from, is to predict that the case in question is at the __59__ of Y. Thus in every case $Y' = \underline{60}$, and the standard error of prediction, $\sqrt{\Sigma(Y - Y')^2/n}$, becomes $\sqrt{\Sigma(Y - \overline{Y})^2/n}$. But the latter expression is the standard deviation of Y, S_Y. So here, with a correlation of zero, the standard error of prediction, __61: symbol?__, is the same as the standard deviation of Y, __62: symbol?__, and there is no reduction in predictive error.

58.
59.
60.
61.
62.

If the correlation rises to .25 (plus or minus), S_{YX} will be somewhat less than S_Y, indicating an improvement in the errors of prediction. But as the table shows, the reduction is trivial, only __63__ %. Even a correlation of .75 (plus or minus) yields a reduction in predictive error of only __64__ %.

63.
64.

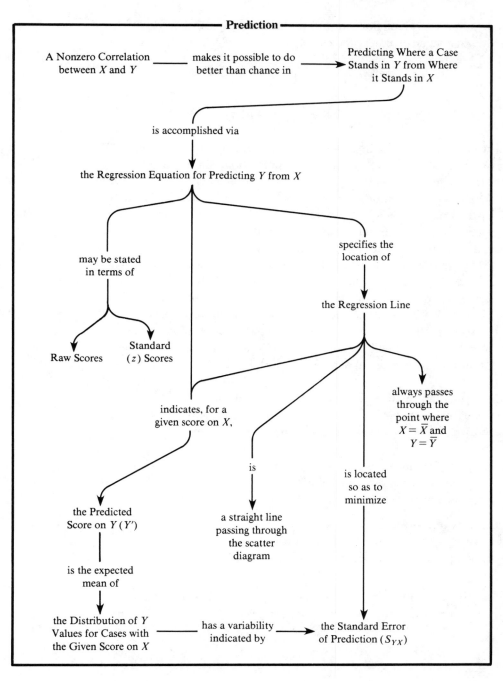

Prediction

A Nonzero Correlation between X and Y ——— makes it possible to do better than chance in ——→ Predicting Where a Case Stands in Y from Where it Stands in X

is accomplished via

the Regression Equation for Predicting Y from X

may be stated in terms of

specifies the location of

the Regression Line

Raw Scores   Standard (z) Scores

indicates, for a given score on X,

always passes through the point where $X = \overline{X}$ and $Y = \overline{Y}$

is

is located so as to minimize

the Predicted Score on Y (Y')

a straight line passing through the scatter diagram

is the expected mean of

the Distribution of Y Values for Cases with the Given Score on X ——— has a variability indicated by ——→ the Standard Error of Prediction (S_{YX})

In general, if our purpose is prediction, we will not achieve any substantial reduction in predictve error unless the correlation is *very* <u>65: strong</u>   65. <u>or weak?</u> .

PATTERNS IN PREDICTIONS

To understand what you are doing in using the regression equation to predict a Y score given an X score, and to check to see if the prediction you are making is a reasonable one, it is helpful to think: where is X in relation to its mean $\overline{X}$? And where is Y predicted to be in relation to *its* mean $\overline{Y}$? Where Y is predicted to be in relation to $\overline{Y}$ depends on: a) where X is in relation to $\overline{X}$, and on b) whether r is positive, zero, or negative. You can figure out for yourself exactly how this works by using the standard-score form of the regression equation.

For example, suppose you want to predict the score on Y for a case whose X value is above the mean of X. Suppose further that the correlation between X and Y is positive. The regression equation says that $z_{Y'} = rz_X$. Here r will be some positive number, while z_X will also be positive, because a raw score above its mean has a positive z score. The product of two positive numbers is also positive; thus the equation predicts that this case will have a positive z score on variable Y. And a positive z score indicates a raw score above the mean of Y. Thus if $X > \overline{X}$ and if r is positive, $Y' > \overline{Y}$.

This generalization is indicated in the upper-left corner of the table below. Using reasoning like that illustrated above, fill in the other cells of the table, in each case putting the symbol $>$, $=$, or $<$ between the Y' and the $\overline{Y}$. Remember:

A raw score above the mean has a positive z score.
A raw score at the mean has a z score of zero.
A raw score below the mean has a negative z score.

The product of two positive numbers is positive.
The product of two negative numbers is positive.
The product of a positive and a negative number is negative.

	If r is positive	If r is zero	If r is negative
If $X > \overline{X}$	$Y' > \overline{Y}$	$Y' \quad \overline{Y}$	$Y' \quad \overline{Y}$
If $X = \overline{X}$	$Y' \quad \overline{Y}$	$Y' \quad \overline{Y}$	$Y' \quad \overline{Y}$
If $X < \overline{X}$	$Y' \quad \overline{Y}$	$Y' \quad \overline{Y}$	$Y' \quad \overline{Y}$

SELF-TEST

Y' is the symbol for ...   A) the mean of Y.   B) a predicted standard score on Y.   C) the standard error of prediction.   D) a predicted raw score on Y.

1.   A   B   C   D

Which can be used to predict a score on Y from knowledge of a score on X?   A) The regression line.   B) The regression equation.   C) Both of these.   D) Neither of these.

2.   A   B   C   D

The error of prediction for a given subject is ...   A) $(Y - Y')$.   B) $(Y' - \bar{Y})$.   C) $(Y - \bar{Y})$.   D) $\Sigma(Y - Y')^2$.

3.   A   B   C   D

Cases with a given score on X will have a variety of scores on Y, and the predicted value of Y for these cases is an estimate of the _____ of their Y values.   A) standard deviation   B) mean   C) variance   D) median

4.   A   B   C   D

No matter what the value of r is, the regression equation predicts that if $X = \bar{X}$, Y' is ...   A) above the mean of Y.   B) at the mean of Y.   C) below the mean of Y.   D) zero.

5.   A   B   C   D

A teacher finds that her pupils' scores on their first arithmetic test are positively but not perfectly correlated with their scores on the second arithmetic test. The student with the top score on the first test was three standard deviations above the mean. When the teacher looks to see how this student did on the second test, she will probably find that the score is ...   A) even more than three standard deviations above the mean.   B) exactly three standard deviations above the mean.   C) less than three standard deviations above the mean, but still above it.   D) below the mean.

6.   A   B   C   D

A sample of 43 scores called X has a mean of 12.2 and a standard deviation of 3.9. Paired with each X score is a score called Y, and the Y values have a mean of 78.3 with a standard deviation of 7.8. The correlation between X and Y is exactly zero. Which is the best prediction of Y for a case whose score on X is 4?   A) Above 78.3.   B) 78.3.   C) Below 78.3.   D) Zero.

7.   A   B   C   D

Suppose the correlation in Question 7 is .47. Now which is the best prediction of Y for a case whose score on X is 4?   A) Above 78.3.   B) 78.3.   C) Below 78.3.   D) Zero.

8.   A   B   C   D

S_{YX} is a kind of ...   A) mean.   B) variance.   C) range.   D) standard deviation.

9.   A   B   C   D

In which case does the regression equation predict that $Y' = \overline{Y}$ for every value of X?   A) $r > 0$ B) $r = 0$ C) $r < 0$ D) In none of these cases.

10.   A   B   C   D

CHAPTER 9

INTRODUCTION TO STATISTICAL INFERENCE AND PROBABILITY

OBJECTIVES

You learned in Ch. 1 that statistical techniques are of two kinds: descriptive and inferential. Chs. 2–8 presented various techniques of description. Ch. 9 now introduces techniques of inference. In studying this chapter, work to understand the following symbols, terms, concepts, principles, and techniques:

Random Sampling
   The Role of Chance in the Formation of a Sample   •   Sampling Variation
The Key Question in Statistical Inference
Probability (Pr)
   Sampling Experiment   •   Definition of Probability   •   Computation
   when Outcomes are Equally Likely
Probability Distribution
The Normal Curve as a Theoretical Probability Distribution
The OR/Addition Rule
   Mutually Exclusive Events
The AND/Multiplication Rule
   Independent Events

Also in studying this chapter, work to learn these skills:

Computation of Probability when Outcomes are Equally Likely
Computation of the Probability of One Event Or Another Or Another ...
Computation of the Probability of One Event And Another And Another ...

REVIEW

Ch. 1 informed you that statistical techniques are of two kinds: descriptive and inferential. The initial chapters of your text introduced you to description, and you learned a variety of techniques for describing a collection of scores: cast it into a frequency distribution, perhaps grouping the scores, perhaps adding a column of cumulative values; graph it as a histogram or frequency polygon or cumulative percentage frequency curve; name its shape; compute a measure of central tendency such as the mean; calculate a measure of variability such as the standard deviation. Ch. 9 now introduces techniques of inference, which permit us to draw proper conclusions from a collection of observations that is only a sample from a larger population.

The Key Question in Statistical Inference

In selecting a sample from a larger population, __1__ factors always play a role. The process of selection called random sampling, for example, is a lottery-type procedure in which chance factors dictate who is to be included in the __2__. The purpose of inferential statistics is to take these chance factors into account in drawing conclusions from a sample of observations. And the key to solving problems of statistical inference is to answer the question: What kind of sample results can we expect on the basis of __3__ alone?

Thus we need to know how chance operates. The study of chance and its effects is known as __4__ theory.

1.

2.

3.

4.

Probability

To speak precisely about probability, the concept of sampling experiment is helpful. A sampling __5__ is something like drawing a card from a well-shuffled deck or flipping a coin. It is a procedure that yields, whenever it occurs, just one of two or more possible results. The possible results are called events. In these terms, the probability of an event is the __6__ of times we would expect the event to occur in an infinitely long series of identical sampling __7__.

5.

6.
7.

What, then, does it mean to say that the probability of getting "heads" in flipping a certain coin is .50? It means that we expect the proportion of heads over an indefinitely long series of flips of that coin to be .50 (one-half). That is, we expect the coin to come up heads half the time *in the long run*. A probability of .50 does not mean that one out of every two flips will be heads. It does not mean that two of every four, or four of every eight, or eight of every sixteen flips will be heads. In the *short* run, over any small number of flips, we could get all heads, all tails, or anything in between. But if we keep flipping and flipping that coin, and if the probability of heads is really .50, the proportion of heads will "average out" to be one-half. In general, a probability is a proportion based on the way things turn out in the long run.

Using this definition of probability is impractical in most instances, because we cannot repeat the sampling experiment often enough to get a good approximation to a long-run proportion. That is, we cannot usually make the run long enough. But for sampling experiments whose possible outcomes are all equally likely, there is a simple rule for calculating a probability. If all the possible outcomes of a sampling experiment are equally likely, the probability of the occurrence of an event is the _8_ of   8.
the possible outcomes characterized by the event.

Consider as an example the drawing of a card from a deck of cards. This is a sampling experiment that yields a sample of size one (one particular card). The possible outcomes of the experiment are the various _9_ that make up the deck. If we draw a card from a standard deck in   9.
such a way that all possible outcomes are equally likely, what is the probability of the event "club"? The definition of probability tells us that the probability of getting a club is the _10_ of times a club will occur in   10.
the long run if we repeat the process of drawing a card from that deck over and over. (The deck must be a full one for every draw.) To determine this probability, we do not have to repeat the sampling experiment over and over; in fact, we don't have to do it even once. Because we are drawing the card in a way that makes all the possible outcomes equally likely, the rule cited above applies. It tells us that the probability of the event "clubs" is the _11_ of the 52 possible outcomes   11.
characterized as a club. Because there are 13 cards that are clubs in a standard deck, the probability is 13/52, or one-quarter (.25).

The equally-likely rule is important in statistics, because it applies to cases of random sampling. By definition, sampling is random if the various possible outcomes of the sampling are equally likely. (Ch. 10 will present this definition.)

Probability Distributions

In Ch. 2, you learned how to display a collection of scores in a frequency distribution, which is a kind of _12: table or graph?_ . And you learned that   12.
a frequency distribution can list not just the raw frequency for each score or each class interval, but also the relative frequency—that is, the proportion or the _13_ . (Look at the table at the top of p. 157 for an   13.
example.) What does this have to do with probability? If we select a case

from the distribution at random, the relative frequencies are the __14__ that   14.
apply to our sampling experiment.

For example, if we select at random one of the 200 students whose
grades are shown in the table on p. 157, by the definition of random
sampling we are selecting a student in such a way that all 200 of them
have the same chance of being selected. (Each has a probability of $1/200$
of being chosen.) What, then, is the probability of getting a student who
earned a grade of B? According to the equally-likely rule, it is the __15__ of   15.
the outcomes characterized by the event B—that is, the proportion of the
students who earned that grade. The table indicates that .30 of the
students earned this grade. Therefore, .30 is the __16__ of selecting by   16.
random draw a student with a grade of B.

In statistics, the most important application of the insight that a
relative frequency distribution can be interpreted as a probability distribution involves the normal curve. As a model for the shape of a
distribution, the normal curve specifies the relative frequencies with
which cases occur in various parts of the distribution. And these relative
frequencies can be viewed as probabilities.

For example, suppose there is a distribution of scores whose shape is
closely modeled by the normal curve. If we draw a score at random from
this distribution, what is the probability that it will be above the mean?
Because we are drawing at random, all scores in the distribution have the
__17__ chance of coming up in our sample of one. Thus the equally-likely   17.
rule applies, and the *probability* of getting a single score whose value is
above the mean is the *proportion* of all the scores that are __18__ .   18.

Now we can make use of the principle that proportion of area under a
frequency curve equals relative frequency of scores. (See pp. 37-39 of the
text for a review of this point.) The principle tells us that the relative
frequency with which scores occur above the mean equals the proportion
of area under the normal curve that falls above the mean. Appendix F
indicates that this proportion is one-half. (The mean has a z value of
__19__ , and the top line of the table in Appendix F says that the area   19.
beyond a z value of zero is .5000.) This proportion of area, one-half, is
also the proportion of __20__ of the kind that concerns us, and that   20.
proportion of scores gives us the probability that we originally asked
about.

To summarize the line of thought here, if we draw a single score at
random from a distribution whose shape is normal, the probability that
the score will lie above the mean can be found by determining the __21__ of   21.
scores that lie above the mean, and this proportion of scores can in turn
be found by determining the proportion of __22__ that is above the mean in   22.

the frequency polygon of a distribution with a normal shape—which is one-half.

The OR/Addition Rule

The equally-likely rule is one of three rules for computing probabilities that Ch. 9 introduces. A second rule tells us how to compute the probability that one event *or* another will occur. This rule applies only if the events are mutually __23__, which means that if one event occurs in a particular instance, the other __24: can or cannot?__ also occur in that instance. For example, in drawing from a deck of cards, the events "face card" and "ten" are __25__ exclusive, because if a face card turns up in a particular draw, a ten __26: can or cannot?__ also occur on that draw, and vice versa. On the other hand, the events "face card" and "club" are not __27__, because if a face card occurs on a give draw, it is still possible that the draw has yielded a club. For example, we could have selected the queen of clubs.

23.

24.

25.

26.

27.

If two events are mutually exclusive, the probability of the occurrence of either one *or* the other is the __28__ of the probability of the first event __29: plus or times?__ the probability of the second. This OR/addition rule can be extended to cases in which we want to know the probability of getting any one of three or more mutually __30__ events. We just __31__ the probabilities of the individual events.

28.

29.

30.

31.

The AND/Multiplication Rule

The third rule introduced in Ch. 9 tells us how to compute the probability that one event *and* another will occur. This rule applies only if the events are __32__, which means that the probability of one event remains the same regardless of whether the other occurs or not. If two events are __33__, the probability that one *and* the other will occur is the __34__ of the probability of the first event __35: plus or times?__ the probability of the second event. This AND/multiplication rule can be extended to cases in which we want to know the probability of getting all of three or more independent events; we just __36: add or multiply?__ all of the individual probabilities together.

32.

33.

34.

35.

36.

The OR/addition and AND/multiplication rules can be used in combination. And the individual probabilities that enter into the computations for either rule alone or both together can be found from the equally-likely rule, if it applies. Thus the three rules introduced in this chapter can be used in any combination.

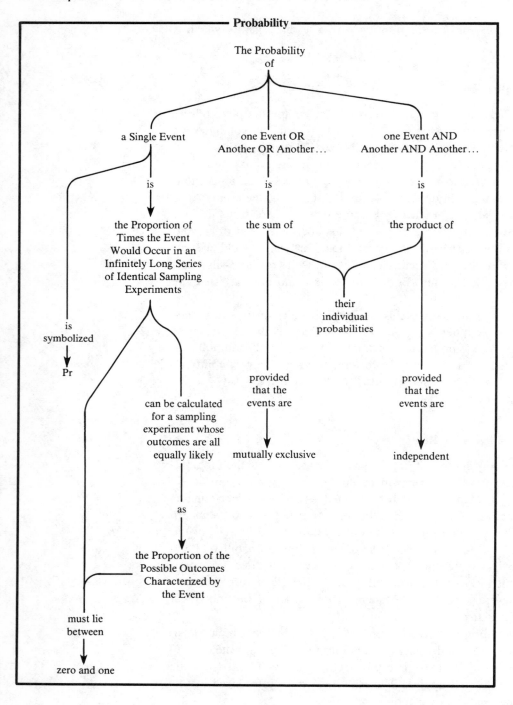

Probability

The Probability of

a Single Event

one Event OR
Another OR Another...

one Event AND
Another AND Another...

is

the Proportion of
Times the Event
Would Occur in an
Infinitely Long Series
of Identical Sampling
Experiments

is

the sum of

is

the product of

their
individual
probabilities

is
symbolized

Pr

can be calculated
for a sampling
experiment whose
outcomes are all
equally likely

provided
that the
events are

mutually exclusive

provided
that the
events are

independent

as

the Proportion of the
Possible Outcomes
Characterized by
the Event

must lie
between

zero and one

SELF-TEST

Which is the key question in statistical inference?   A) What is the mean of my sample?   B) What is the standard deviation of my sample?   C) To what extent did chance factors play a result in producing my sample?   D) What kind of sample results should I have expected on the basis of chance alone?

1.   A   B   C   D

The probability of rolling "snake eyes" with a pair of honest dice (the probability of getting one spot showing on each cube) is 1/36. What does this number 1/36 mean?   A) Out of every 36 rolls, snake eyes will come up exactly once.   B) Out of every 36 rolls, snake eyes will come up at least once, and maybe more often.   C) Out of an infinite number of rolls, snake eyes will come up exactly 1/36 of the time.   D) Out of an infinite number of rolls, snake eyes will come up at least 1/36 of the time, and maybe more often.

2.   A   B   C   D

Which the following can*not* be a probability?   A) 1.5   B) 1.0 C) 0.5   D) 0.0001

3.   A   B   C   D

If we draw a card at random from a standard deck of 52 playing cards, what is the probability that it will be one of the four aces in the deck? A) 1/52   B) 4/52   C) 1/4   D) 52/4

4.   A   B   C   D

Which of the following may be interpreted as a probability distribution? A) A relative frequency distribution.   B) The normal curve.   C) Both of these.   D) Neither of these.

5.   A   B   C   D

Two events are mutually exclusive if ...   A) both can occur at the same time.   B) both cannot occur at the same time.   C) the occurrence of one changes the probability of the occurrence of the other.   D) the occurrence of one does not change the probability of the occurrence of the other.

6.   A   B   C   D

In selecting a card from a standard deck, which two events are mutually exclusive?   A) Drawing an ace and drawing a spade.   B) Drawing a jack and drawing a face card.   C) Drawing a red card and drawing a three.   D) Drawing a king and drawing a queen.

7.   A   B   C   D

The AND/multiplication rule applies only to events that are ... A) independent.   B) mutually exclusive.   C) only weakly correlated. D) dependent.

8.   A   B   C   D

Suppose we have three standard decks of playing cards, and we draw a card at random from each. What is the probability of obtaining an ace on

9.   A   B   C   D

each draw?   A) $1/4 + 1/4 + 1/4$ B) $1/4 \times 1/4 \times 1/4$ C) $4/52 + 4/52 + 4/52$ D) $4/52 \times 4/52 \times 4/52$

Suppose there are two events that are both mutually exclusive and independent. Suppose further that the probability of each is between .1 and .9. How does the probability that the first event and the second event will both happen compare with the probability that either the first or the second will happen?   A) The first probability is larger.   B) The two probabilities are equal.   C) The first probability is smaller.   D) It depends on the exact numbers involved.

10.   A   B   C   D

CHAPTER 10

RANDOM SAMPLING DISTRIBUTIONS

OBJECTIVES

As you learned in Ch. 9, the key question in statistical inference is: What kind of sample results can we expect on the basis of chance alone? Ch. 9 presented the basics of the theory of probability, which permits us to deal with chance in a precise way. Ch. 10 now offers the fundamental concepts and principles necessary to understand specifically what happens when we draw a sample from a population.

In studying Ch. 10, work to understand the following symbols, terms, concepts, principles, and techniques:

Models Underlying Statistical Inference
   The Random Sampling Model   •   The Randomization Model
Population
   Parameter
      μ • σ
Sample
   Statistic
Sampling
   Sampling at Random   •   Casual Sampling
The Key Questions in Statistical Inference
Random Sampling Distribution of Means
   Mean ($\mu_{\bar{X}}$)
   Standard Error of the Mean ($\sigma_{\bar{X}}$)
   Shape
      Central Limit Theorem
A Sampling Distribution is a Probability Distribution
There is a Random Sampling Distribution for any Statistic

Also in studying Ch. 10, work to learn these skills:

Determination of the Characteristics of a Random Sampling Distribution of Means

Mean $(\mu_{\bar{X}})$ • Standard Error $(\sigma_{\bar{X}})$ • Shape

Computation of the Probability of Obtaining a Sample Mean Falling in a Given Interval

Computation of a Sample Mean Possessing a Given Probability of Occurrence

REVIEW

Random Sampling and Randomization

There are two models in common use among researchers who employ traditional procedures for statistical inference. One is the random __1__ model, in which the researcher assumes that the observations on hand are a sample randomly selected from a __2__ . Under this model, the researcher's task is to infer the unknown characteristics of the __3__ from the known characteristics of the __4__

The second model is the __5__ model. In this model, the researcher takes a group of subjects who were *not* randomly selected but were merely available, divides them at random into two or more subgroups, and treats each subgroup in a different way. Under this model, the researcher's task is to infer whether the varying treatment made any difference in the subjects' scores beyond the differences due to the chance factors that operated in the formation of the subgroups.

Strictly speaking, the proper statistical procedures for inference are __6: the same or different?__ in these two models. But in most situations in which subgroups are formed by randomization, the random sampling model will lead to statistical conclusions the same as those resulting from the randomization model. In the behavioral sciences, researchers thus almost invariably apply the random __7__ model, which is simpler, even when they use __8__ to form subgroups. Your text accordingly develops and applies only the random sampling model. This model is useful even for comparisons between groups of available subjects that were formed without any sort of random procedure—for example, when we compare one intact group such as a classroom with another. (We might want to compare a class of students using a new text with a class using an old text. Even though students were not assigned to one classroom or the other at random, the random sampling model will still be helpful.)

1.

2.
3.
4.
5.

6.

7.
8.

Populations and Samples

To apply the random sampling _9_ of statistical inference, we must be
clear about the concepts of population and sample. A population consists
of the _10_ set of observations about which we would like to draw
conclusions; a sample is a _11_ of a population.

The characteristics of a population are called _12_ , whereas the characteristics of a sample are called _13_ . To distinguish parameters from
statistics, we use different symbols. The mean of a population of scores is
symbolized _14_ ("mew"), and the standard deviation of a population of
scores is symbolized _15_ ("sigma"). (μ and σ are lower-case Greek
letters.) In contrast, as you have learned, the mean of a sample is
symbolized _16_ , and the standard deviation of a sample is symbolized
_17_ .

9.

10.

11.

12.

13.

14.

15.

16.

17.

Sampling at Random

To apply the random sampling model of statistical inference, we must
also be clear about what random sampling is. A random sample is a
sample of a given size drawn from a population in such a way that each
possible sample of that size from that population has _18: the same
or a different?_ probability of being selected. When a sample is selected in
this way, it is necessarily true that every observation in the population
has an equal opportunity of being included in the _19_ . If some observations had a greater chance of appearing in the sample than others did
(that is, if sampling was biased), then the sample cannot be random.

Can we tell just by looking at a sample whether it was chosen from its
population at random? _20: Yes or No?_ It is how we select the sample,
not the results that we obtain, that determines whether or not a sample is
_21_ . Characteristics of random samples do _tend_ to resemble those of the
parent population, but the characteristics of a particular random sample
may not. So inspecting the sample to see how closely it resembles its
parent population will not tell us whether it was chosen at _22_ .

It is especially important to note that a casual sample is not a random
one. A casual sample is one that an investigator selects haphazardly, the
way a reporter typically chooses subjects for a series of "person-in-the-
street" interviews. Because anyone who chooses a sample in such a casual
way is sure to have some preferences, some of the various subjects who
could contribute observations to the sample are more likely to do so than
others. Thus the resulting sample is not selected at _23_ , and it is very
hazardous to generalize from this sample to the population.

18.

19.

20.

21.

22.

23.

The Random Sampling Distribution of Means

The basic concept in statistical inference is the sampling distribution. Ch.
10 concentrates on one particular kind of sampling distribution, that of
sample means. Imagine a population of scores. Imagine a sampling
experiment in which we select a sample of those scores at random. We
compute the mean of the sample, __24: symbol?__ . Now we repeat the   24.
sampling experiment, drawing another sample of the same size at random
again from the same population. The scores constituting this second
sample will almost certainly differ from the scores constituting the first
sample, and when we compute the mean of the second sample, we will
almost certainly find that it will differ from the mean of the first sample.
We continue this procedure forever, generating an infinitely large collection of means of samples of that given size, samples each drawn at
random from that given population. The collection of sample means is
the __25__ sampling __26__ of means for this particular population and this   25.
particular sample size.   26.
   More formally, a random __27__ distribution of means is the relative __28__   27.
distribution of sample means obtained from an unlimited series of   28.
identical sampling experiments, each experiment yielding a sample of the
same particular size drawn at random from a given population.
   In practice, of course, it is impossible to generate a __29__ sampling   29.
distribution. But mathematicians have discovered principles that enable
us to know what its properties would be. The properties depend on the
characteristics of the population sampled and on the size of the samples.

Characteristics of a Random Sampling Distribution of Means

A complete description of any distribution of numbers includes its mean,
its standard __30__ , and its __31__ . Thus a complete description of a random   30.
sampling distribution of __32__ will specify these three characteristics. They   31.
are treated below in order.   32.
   1. *The mean of a random sampling distribution of means:* This quantity
is symbolized __33__ ("mew sub eks-bar"). (μ is the symbol for the mean of   33.
a __34__ , and indeed, the collection of sample means that constitutes a   34.
random sampling distribution *is* a population, for those are the means of
all possible samples of a given size that can be drawn from a given
population.) In all cases, whatever the standard deviation of the population is, whatever the shape of the population is, whatever the sample size
is, the mean of a random sampling distribution of means, __35: symbol?__ ,   35.
is the same as the mean of the population of raw scores, __36: symbol?__ .   36.
That is, $\mu_{\bar{X}} = \mu$.

2. *The standard deviation of a random sampling distribution of means:* This quantity is called the standard __37__ of the mean, and it is symbolized __38__ ("sigma sub eks bar"). (The σ is appropriate here, because it is the symbol for the standard deviation of a population, and a random sampling distribution is a population, as noted just above.) The standard error of the __39__ depends only on the standard __40__ of the population, σ, and on the sample size, *n*. The formula is $\sigma_{\bar{X}} = $ __41__ . This formula tells us that for every different value of *n*, there is a different sampling distribution of means from a given population, because changing *n* changes the standard deviation (the "standard error") of the sampling distribution. The formula also tells us that if sample size, *n*, becomes larger, there will be __42: more or less?__ variation from one sample mean to another within the sampling distribution. The formula tells us further that with greater variability among the raw scores in the population, as measured by their standard deviation, σ, there will be __43: more or less?__ variability among the means in the random sampling distribution.

37.
38.

39.
40.
41.

42.

43.

3. *The shape of a random sampling distribution of means:* Two principles apply here. The first principle says that if the population of scores is normally distributed, the sampling distribution of means will also be __44__ distributed. The second applies to populations that are not normally distributed, and it assures us that a random sampling distribution of means *tends* toward a normal distribution irrespective of the shape of the population of scores sampled, with the approximation to the normal shape becoming increasingly close as sample size __45: increases or decreases?__ . The latter principle is known as the Central __46__ Theorem. How large must sample size be for the sampling distribution of means to be treatable as normally distributed? About 25 to 100.

44.

45.
46.

Putting the Random Sampling Distribution of Means to Use

The principles specifying the mean, standard deviation, and shape of a random __47__ distribution of means can be put to good use to solve a variety of important problems. The problems involve the question of how typical or how rare a particular sample mean is. If the sample was chosen at random from a population whose characteristics are known, we can determine the characteristics of the random sampling __48__ of means of samples of the kind in question, and we can then calculate helpful probability values. Section 10.7 supplies details and examples.

47.

48.

In working problems of this kind, it is necessary to state the location of a sample mean within a random sampling distribution of means, and this is best done in terms of __49__ scores. The general formula for a *z* score is

49.

(score − _50_)/ _51_ . In working with a sample mean, the sample mean, $\overline{X}$, serves as the _52_ in this formula; the mean of the population of raw scores, μ, serves as the _53_ in the formula; and the standard _54_ of the mean serves as the standard deviation. Therefore the z score that describes the location of a mean within the sampling distribution of sample means of its kind has the formula $z = $ _55_ .

50.

51.

52.

53.

54.

55.

Generality of the Concept of Sampling Distribution

The notion of sampling distribution applies to _56: all or only some?_ sample statistics. For example, there are sampling distributions of sample medians, and there are sampling distributions of sample standard deviations. In general, a random _57_ distribution of a statistic is the _58_ frequency distribution of that statistic obtained from an unlimited series of identical sampling experiments in which a sample of some particular size is repeatedly drawn at random from a given _59_ .

56.

57.

58.

59.

MNEMONIC TIP

The text has now informed you that a characteristic of a population is called a parameter, while a characteristic of a sample is called a statistic. You can easily remember the distinction between the terms for a characteristic if you note that *population* and *parameter* both begin with a *P*, while *sample* and *statistic* both begin with an *S*.

SAY IT RIGHT

Ch. 10 of your text has introduced the lower-case Greek letter μ as the symbol for the mean of a population. Be sure you read and think of this as "mu" (pronounced "mew"). It's not the Roman letter u, and it's not pronounced "you." The μ is the Greek letter corresponding to the Roman m, and it would be the initial letter in the Greek spelling of *mean*.

Also on this topic of pronunciation, remember that the capital Greek sigma, Σ, is to be read as "the sum of." ΣX is "the sum of X," not "ee eks." (But the lower-case sigma, σ, is to be read "sigma," as this chapter also informed you. It would be the initial letter in the Greek spelling of *standard deviation*, and it's thus an appropriate symbol for the standard deviation of a population.)

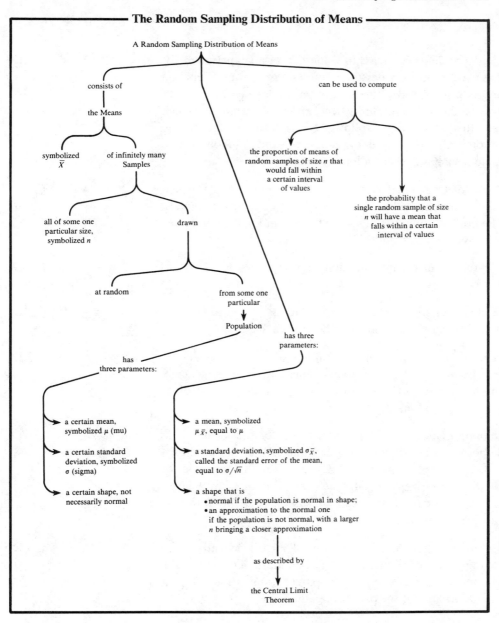

The Random Sampling Distribution of Means

A Random Sampling Distribution of Means

consists of

the Means

symbolized
$\overline{X}$

of infinitely many
Samples

can be used to compute

the proportion of means of
random samples of size n that
would fall within
a certain interval
of values

the probability that a
single random sample of size
n will have a mean that
falls within a certain
interval of values

all of some one
particular size,
symbolized n

drawn

at random

from some one
particular

Population

has three
parameters:

has
three parameters:

a certain mean,
symbolized μ (mu)

a certain standard
deviation, symbolized
σ (sigma)

a certain shape, not
necessarily normal

a mean, symbolized
$\mu_{\overline{X}}$, equal to μ

a standard deviation, symbolized $\sigma_{\overline{X}}$,
called the standard error of the mean,
equal to $\sigma/\sqrt{n}$

a shape that is
• normal if the population is normal in shape;
• an approximation to the normal one
  if the population is not normal, with a larger
  n bringing a closer approximation

as described by

the Central Limit
Theorem

SELF-TEST

A psychologist asks students in a class to volunteer for an experiment; 20 of them do. The psychologist divides them into two groups by flipping a coin, tests one group for speed of learning when under the influence of caffeine, and tests the other group when not so influenced. Which procedure has this researcher followed? A) Random sampling. B) Randomization. C) Both random sampling and randomization. D) Neither random sampling nor randomization.

1.   A   B   C   D

If the psychologist of Question 1 is a typical behavioral scientist, which model of statistical inference will she or he apply to the speed-of-learning scores? A) The random sampling model. B) The randomization model. C) Both models.   D) Neither model.

2.   A   B   C   D

The symbol for the mean of a population of raw scores is ...   A) μ. B) $\mu_{\bar{X}}$.   C) σ.   D) $\bar{X}$.

3.   A   B   C   D

A statistic is a characteristic of a ...   A) population.   B) random sampling distribution of means.   C) sample.   D) normal curve.

4.   A   B   C   D

Which statement is true?   A) A random sample always closely resembles its parent population.   B) A random sample may differ drastically from its parent population.   C) A random sample must consist of at least 25 scores.   D) A random sample is always normal in shape.

5.   A   B   C   D

A random sampling distribution of means consists of the means of ... A) all possible populations.   B) all possible samples that can be drawn at random from a particular population.   C) an infinitely long series of samples, all of the same size, and all drawn at random from a given population.   D) n different samples, all of size n, and all drawn at random from a given population.

6.   A   B   C   D

The mean of a random sampling distribution of means depends on ... A) the mean of the parent population.   B) the standard deviation of the parent population. C) the shape of the parent population.   D) all three of these characteristics of the parent population.

7.   A   B   C   D

The standard deviation of a random sampling distribution of means depends on ...   A) the mean of the parent population.   B) the standard deviation of the parent population.   C) the shape of the parent population.   D) all three of these characteristics of the parent population.

8.   A   B   C   D

The standard error of a random sampling distribution of means depends on ...   A) the standard deviation of the parent population.   B) the size of the samples.   C) both of these.   D) neither of these.

9.   A   B   C   D

The z score that specifies the location of a sample mean within a random sampling distribution of means is calculated by computing $(\overline{X} - \mu)$ divided by what?   A) $\mu_{\overline{X}}$ B) σ C) S D) $\sigma_{\overline{X}}$.

10.   A   B   C   D

CHAPTER 11

INTRODUCTION TO HYPOTHESIS TESTING: I

OBJECTIVES

Ch. 9 introduced the basic concepts of probability, and Ch. 10 presented the important facts about random sampling distributions of means. Ch. 11 now shows you how to put all of this information to practical use in testing hypotheses about the mean of a population. In studying Ch. 11, work to understand the following symbols, terms, concepts, principles, and techniques:

The Logic of Hypothesis Testing
   The Level of Significance (α)
Estimation of the Standard Error of the Mean ($\sigma_{\bar{X}}$)
   The Bias in S as an Estimate of σ • s • $s_{\bar{X}}$
Null Hypothesis (H_0)
Alternative Hypothesis (H_A)
   Nondirectional (Two-Tailed)   •   Directional (One-tailed)   •   Choice between Nondirectional and Directional
Locating $\bar{X}$ in the Random Sampling Distribution of Means Implied by the Null Hypothesis
   t as an Approximate z
Region of Acceptance
Region of Rejection
   Critical Value(s)
      The Effect of Sample Size on the Critical Value(s)
Acceptance of H_0
Statistical Significance
   Statistical Significance with a Large Sample May be Insignificant
The Nature and Role of the Null and Alternative Hypotheses
   The Association of the Research Hypothesis with H_A •   The Reason for Testing H_0 Rather than H_A

Also in studying Ch. 11, work to learn these skills:

Selection of the Null Hypothesis for a Given Investigation
Selection of the Alternative Hypothesis for a Given Investigation
Computation of an Estimate of the Standard Deviation of a Population, Given a Sample of Scores
Computation of an Estimate of the Standard Error of the Mean, Given a Sample of Scores
Execution of a Test of a Null Hypothesis about the Mean of a Population, Given a Large Sample
Computation of the t Ratio   •   Computation of the Critical Value(s) when Sample Size is Large   •   Decision on H_0

REVIEW

Estimating the Population's Standard Deviation and the Standard Error of the Mean

To test a hypothesis about the mean of a population, it is necessary to calculate the standard deviation of a random sampling distribution of means of samples drawn from that population. The standard deviation of a sampling distribution of means is called the standard _1_ of the mean, and it is symbolized _2_ . On p. 178, you learned that $\sigma_{\bar{X}}$ is equal to _3_ . This formula requires knowledge of σ, the _4_ of the population, but in actual practice σ will be unknown, and we must estimate it from a sample. You might think that the standard deviation of the sample, _5: symbol?_ , would give the best estimate, but it proves to be slightly too _6: small or large?_ , on the average.

    A statistic called s ("little es") provides a better estimate of σ. Its formula is:
$$s = 7$$
The formula that defines S (which you should now read as "big es") is _8_ . Compare the two formulas, and you will see that s differs only in that the divisor is _9_ rather than n. This change makes s slightly _10: smaller or larger?_ than S.*

    Substituting s for σ in the formula for the _11_ error of the _12_ yields the working formula for estimating the standard error. The estimate of

1.
2.
3.
4.

5.
6.

7.

8.
9.
10.

11.
12.

*It is now highly important to distinguish big S from little s. In your writing, I suggest that you do it by making the capital letter a large version of the printed variety and the lower-case letter a small version of the script variety, like these: **S𝒜**

the standard error is symbolized __13__ , to distinguish it from $\sigma_{\bar{X}}$, which is   13.
the real thing. The formula for the estimate is:

$$s_{\bar{X}} = 14$$ 14.

Stating Hypotheses

In testing an hypothesis about the mean of a population, a researcher
states that the mean has a certain value (for example, $\mu = 126$). Such a
statement is called a __15__ hypothesis, and its symbol is __16__ ("aitch   15.
null"). This is the hypothesis that the researcher will decide to accept or   16.
__17__ .   17.

The researcher must also state an alternative hypothesis, symbolized
__18__ ("aitch sub ay"). This alternative hypothesis may be nondirectional   18.
or directional. A *non*directional hypothesis states that the population
mean does not equal the value specified by the __19__ hypothesis (for   19.
example, $\mu \neq 126$), without specifying whether the mean is less than or
greater than the value stated by the null __20__ . Use of a nondirectional   20.
alternative allows the investigator to reject the __21__ hypothesis if the   21.
evidence points with sufficient strength either to the possibility that μ is
greater than the value hypothesized by H_0 *or* to the possibility that it is
less.

A directional __22__ hypothesis takes one of two forms. It states either   22.
than the population __23__ is less than the value specified by the null   23.
hypothesis (for example, $\mu < 126$) or that the population mean is __24__   24.
than this value ($\mu > 126$). A directional alternative hypothesis is appropriate when a researcher is interested in learning *only* whether the true
value of μ differs from the hypothesized value (the value specified by the
__25__ hypothesis) *in a particular direction.*   25.

The choice between a directional and a nondirectional __26__ hypothesis   26.
should be determined by the rationale that gives rise to the study, and it
should be made __27: before or after?__ the data are gathered.   27.

When we state a *non*directional alternative hypothesis, we call the
resulting test a __28: one-tailed or two-tailed?__ test, because we will reject H_0 28.
if the obtained sample mean falls in an extreme position in __29: just one__   29.
__or either?__ tail of the sampling distribution of means. See Figures 11.4
and 11.5 on pp. 199 and 200 for illustrations.

When we state a directional __30__ hypothesis, we call the resulting test   30.
__31: one-tailed or two-tailed?__ , because we will reject __32: symbol?__ if the   31.
obtained sample mean falls in an extreme position in just one tail of the   32.
sampling distribution of means. See Figure 11.6 on p. 203.

The Level of Significance

In testing a null hypothesis about the mean of a population, we draw a sample at random from that population, and we determine the mean of the sample. If the sample mean is so different from what we would expect to find if H_0 were true that its appearance would be unlikely if H_0 were true, we will   33: accept or reject?   the hypothesis. But what degree of rarity of occurrence is so great that it is better to reject the null hypothesis than to accept it? It is common practice to reject H_0 if the sample mean is so deviant that its probability of occurrence in random sampling when H_0 is true is .05 or less—or alternatively,   34   or less. Whatever number we choose, it serves as a criterion for deciding whether a sample result would be so rare if   35   were true that we are justified in rejecting the null. This criterion is called the level of   36   for our test. Its symbol is the lower-case Greek letter alpha: α.

33.

34.

35.

36.

Picturing the Sampling Distribution Implied by H_0

What sample means would occur if H_0 were true? If it were true, the random sampling distribution of means (for whatever sample size we use) would center on the value specified by the   37   hypothesis. Why? Because the mean of a random sampling distribution of means,   38: symbol? , is equal to the mean of the population of raw scores, μ. That is, $\mu_{\bar{X}} = \mu$. The null hypothesis specifies a certain value for μ, a value symbolized   39   ("mew hype"), so if the null hypothesis is true, $\mu_{\bar{X}} = \mu_{\text{hyp}}$.

37.

38.

39.

We can thus draw a picture of the random sampling distribution of means (for whatever sample size we use) on the assumption that the   40   hypothesis is true. Such a picture appears on pp. 193, 194, 198, 199, 200, and 203 in your text. Each pictured distribution centers on μ_{hyp}, which takes the value   41   for the birthweight example used in the chapter.

40.

41.

Regions of Acceptance and Rejection

The sampling distribution of   42   that would occur if H_0 were true is divided into a region of acceptance and one or two regions of   43 . If the obtained sample mean falls within a region of acceptance, we accept the   44   hypothesis. If the obtained sample mean falls within a region of rejection, we   45   the null hypothesis.

42.

43.

44.

45.

The region of   46   always covers the central portion of the distribution and always includes μ_{hyp}, as the figures in the text show.

46.

If the alternative hypothesis is *non*directional, there are two regions of rejection, one in each tail. See the figures on pp. 194, 199, and 200 for examples. If the alternative hypothesis is directional, there is just one region of __47__, located in just one of the tails. If the alternative hypothesis specifies a value for μ that is *less* than that named by the null, the region of rejection lies in the left (the __48: upper or lower?__) tail, as in the left half of the figure on p. 203. If the alternative hypothesis specifies a value for μ that is *greater* than that named by the null, the region of rejection lies in the __49: left or right?__ (the upper) tail, as in the right half of the figure on p. 203.

47.

48.

49.

In every case, the total area of the region or regions of rejection (when it is expressed as a proportion of the entire area under the curve) is equal to α, the level of __50__. If there are two regions, each has half of this amount, as in the figures on pp. 194, 199, and 200.

50.

Using *t* Ratios in Hypothesis Testing

In this chapter, the X axis in the pictures of the random sampling distribution of means implied by the null hypothesis is divided into regions of acceptance and rejection by one or two z scores. These z scores are called __51__ values, symbolized z_{crit}. If we know the level of significance, __52: symbol?__ , and if we know the alternative hypothesis, __53: symbol?__ , we can figure out the critical value(s) by consulting the table of areas under the normal curve in Appendix F. Using the normal curve in this way is not quite right, and the next chapter will explain how to do things properly, but the normal curve will be good enough for samples of over 30 or 40 cases.

51.

52.

53.

To determine whether the obtained sample mean falls within the region of acceptance or the region of __54__, it is desirable to express the sample mean as a z score. In general, a z score has the form $z = $ (score $-$ __55__)/standard deviation. In a sampling distribution of means, the obtained mean serves as the score in this formula, the population mean serves as the __56__ in the formula, and the standard __57__ of the mean serves as the standard deviation. (This point was introduced in Ch. 10 on pp. 181-182.) The formula for the z score that expresses the location of a sample mean within a sampling distribution of means is thus:

54.

55.

56.

57.

$$z = (\overline{X} - \mu)/\sigma_{\overline{X}}$$

Now since we are working with the sampling distribution that would obtain if the null hypothesis were true, we will take the value of the population mean, μ, to be the one specified by the null hypothesis, μ_{hyp}. And since we do not know the standard error of the mean, $\sigma_{\bar{X}}$, we will substitue for it the estimate __58: symbol?__ . Thus the best we can do to express the location of our obtained sample mean within the sampling distribution of means that would obtain if the null hypothesis were true is to compute the following:

$$t = \left(\bar{X} - \mu_{hyp} \right)/s_{\bar{X}}$$

58.

This quantity is not a true z score and is instead called a t ratio. It is not a true z score because we have had to substitute an estimate for the real thing in the denominator (the bottom) of the ratio. The t ratio serves as an approximate z score.

Concluding a Test

To conclude the test of a null hypothesis about the mean of a population, as the test is introduced in this chapter, we compare our t ratio, which expresses the approximate location of the obtained sample __59__ within the random sampling distribution of means implied by the null hypothesis, with the critical value or values of z. If the obtained sample mean, as indicated by the t, falls in the region of rejection, we reject the __60__ hypothesis. If the obtained sample mean, as indicated by the t, falls in the region of acceptance, we accept the __61__ hypothesis.

59.

60.

61.

We accept H_0 only when we do not have sufficient evidence to reject it —only, that is, when the obtained sample mean is not one that would be rare if H_0 were true. The decision to accept H_0 thus does not mean that H_0 must be true. Accepting H_0 does not mean even that H_0 is likely to be true. It means only that there was not enough evidence to __62__ it, so that it *could* be true.

62.

If we must accept H_0, then, the test of our hypothesis was inconclusive. It is only when we can __63__ H_0 that we learn something definite from our research.

63.

Fortunately, though, in doing research we usually *want* to reject whatever null hypothesis we state. Why? Because the substantive question that motivates an investigation is usually expressed more directly in the __64: null or alternative?__ hypothesis than in the __65: null or alternative?__. Why, then, do we test the null hypothesis rather than the alternative? Because the null states a specific value for the population mean, whereas the alternative does not, and we must have a specific value in order to locate the relevant random __66__ distribution of means.

64.

65.

66.

Basic Elements of the Procedure for Testing a Hypothesis about a Single Mean

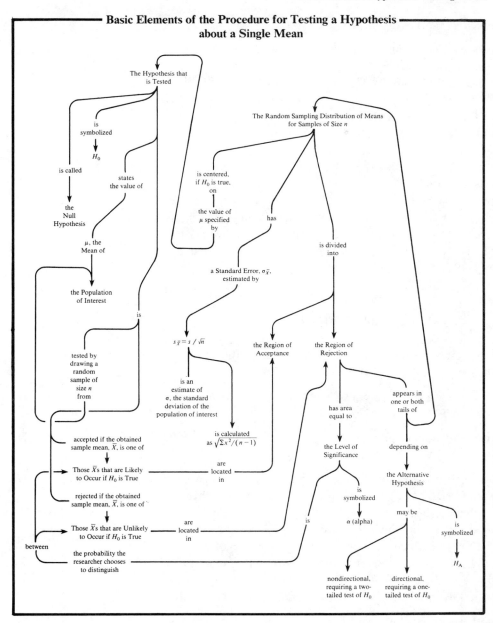

The Terms *Significant and Nonsignificant*

A researcher whose sample results permit the rejection of a null hypothesis describes the results as "statistically significant." If the results do not permit rejection of a null hypothesis, forcing the inconclusive acceptance of the null, the researcher describes the results as "nonsignificant." To understand exactly what the researcher is saying, we must know the level of __67__ that she or he employed in testing the null hypothesis, of course.   67.

The use of the word *significant* in connection with statistical outcomes is unfortunate. In common English the word implies "important," "consequential," or "substantial," but in statistics it means only that the sample results permit the rejection of a null hypothesis at a certain level of significance. Whether the difference between what the null hypothesis said and what is true is large enough to be important is another matter. In fact, very __68: large or small?__ samples may produce results that are   68.
statistically significant but of no practical importance.

SELF-TEST

The symbol for the estimate of the standard deviation of the sampling distribution of means is ...   A) σ B) S.   C) $\sigma_{\bar{X}}$.   D) $s_{\bar{X}}$.

1.   A   B   C   D

Which of these could be a null hypothesis?   A) $\mu < 100$.   B) $\mu = 100$. C) $\mu > 100$.   D) $\bar{X} = 100$.

2.   A   B   C   D

Which is a nondirectional alternative hypothesis?   A) $\bar{X} \neq 100$. B) $\mu \neq 100$.   C) $\bar{X} < 100$.   D) $\mu > 100$.

3.   A   B   C   D

A researcher is comparing a new method of therapy with the treatment that is currently the best. The researcher is interested only in whether the new method is superior; he does not care whether he can distinguish between the state of affairs in which the new method is equal to the old and the state of affairs in which the new method is worse than the old. Which of the following should the researcher use?   A) A directional alternative hypothesis.   B) A nondirectional alternative hypothesis. C) A directional null hypothesis.   D) A nondirectional null hypothesis.

4.   A   B   C   D

In testing a hypothesis about the mean of a population, we assume (for the sake of testing the assumption) that the random sampling distribution of means is centered ...   A) on $\bar{X}$.   B) on the value stated in the alternative hypothesis.   C) on the value stated in the null hypothesis. D) within the range stated by the alternative hypothesis.

5.   A   B   C   D

The region of acceptance always falls where in the hypothesized random sampling distribution of means?   A) In the central portion.   B) In the left tail.   C) In the right tail.   D) In one tail or the other or in both tails, depending on what H_A says.

6.   A   B   C   D

The area of the region of acceptance (expressed as a proportion of the total area under the curve) is ...   A) α.   B) $\alpha/2$.   C) $1 - \alpha$.   D) $1 - \alpha/2$.

7.   A   B   C   D

The general formula for a z score is (score minus what?) divided by standard deviation.   A) predicted score   B) standard error   C) mean   D) null score

8.   A   B   C   D

A t ratio is an approximation to a ...   A) mean.   B) standard error.   C) normal distribution.   D) standard score.

9.   A   B   C   D

Calling the outcome of an investigation statistically significant means that the researcher ...   A) discovered evidence supporting the substantive hypothesis.   B) accepted a null hypothesis.   C) had to raise the level of significance in the course of the study.   D) rejected a null hypothesis.

10.   A   B   C   D

CHAPTER 12

INTRODUCTION TO HYPOTHESIS TESTING: II

OBJECTIVES

Ch. 12 continues your introduction to hypothesis testing by providing additional details. In studying the chapter, work to understand these symbols, terms, concepts, principles, and techniques:

The Difference between a t Ratio and a True z
  The Effect of Sample Size
Degrees of Freedom (df)
  Degrees of Freedom for s and $s_{\bar{X}}$
Student's Distribution of t
  As a Family of Distributions
  The Effect of the Degrees of Freedom
  Appendix G as a Source of Critical Values
  The Assumption of Population Normality
    The Effect of Sample Size
p-Value
  Its Role in Hypothesis Testing • Landmark Values • Difference between a p-Value and a Level of Significance
Error in Hypothesis Testing
  Type I Error
    The Probability of a Type I Error when the Null Hypothesis is True
  Type II Error
    $\alpha = .05$ and $\alpha = .01$ as Compromises

Also in studying Ch. 12, work to learn these new skills (which supplement those taught in Ch. 11):

Computation of the Degrees of Freedom Available in a Single Sample for the Estimates s and $s_{\bar{X}}$

Use of Appendix G to find a Critical Value of t, Given the Applicable Degrees of Freedom

Reporting of a p-Value Relative to the Landmark Values, Given a Judgment of the Statistical Significance

REVIEW

t as an Approximation to z

As you have learned, in testing a null hypothesis about the mean of a _1_, we draw a sample at random from the population and ask a question about the _2_ of our sample: Is this mean, $\overline{X}$, one that is *likely* to occur if the population really had the mean, μ_{hyp} specified in the null hypothesis? Or is this sample mean one that would be *rare* if the population mean were what the null hypothesis states it to be? (If the answer is "likely," we _3_ the hypothesis; if "rare," we _4_ the hypothesis.)

To determine whether the obtained sample mean would be a likely event or a rare one, should the null hypothesis be true, we compare the sample mean with those that would occur if the null were _5: true or false?_. The means that would occur if the null were true are the means in a certain random sampling distribution of _6_ —namely the sampling distribution of means for samples of whatever size we drew and for a population whose mean, μ, is whatever the null hypothesis states it to be.

To locate our obtained sample mean, _7: symbol?_, within this random sampling distribution, we would *like* to compute a z score. The general formula for a z score is:

$$z = (\text{score} - \text{mean})/\text{standard deviation}$$

In our case, we would have:

$$z = \left(\overline{X} - \mu_{hyp} \right)/\sigma_{\overline{X}}$$

In this formula, $\overline{X}$ is the obtained mean of the _8: sample or population?_, μ_{hyp} is the value that the _9_ hypothesis specifies for the population mean, and $\sigma_{\overline{X}}$ is the standard deviation of the sampling distribution of means, which is called the standard _10_ of the mean.

We cannot, however, fill in the exact value of _11: symbol?_ in that formula. The best we can do is to estimate it, using the quantity called _12_, which we can calculate from our sample. When we use this estimate for the denominator (the bottom) of the fraction, we are calculating only an approximation to the z score; it is called a _13_ ratio. The formula

1.

2.

3.
4.

5.

6.

7.

8.
9.

10.

11.

12.

13.

thus becomes:

$$t = \left(\overline{X} - \mu_{hyp} \right)/s_{\overline{X}}$$

Because the statistician who did the pioneering work on this matter published under the pseudonym "Student," the quantity is sometimes called Student's t.

The Effect of Sample Size

If sample size is large (over 30 or 40), the value of $s_{\overline{X}}$ tends to be a good approximation to __14__ . In such a case t will then be a good approximation to __15__ , and we will not go far wrong if we use the table of areas under the normal curve to obtain critical values. This is the procedure that Ch. 11 presented. (See the figures on pp. 199, 200, and 203 for illustrations showing critical values derived from the normal curve.)

But when sample size is small (less than 30 or 40), treating a t ratio as though it were a true __16__ score will produce substantial error. Here we must obtain the critical values from something called the family of t distributions.

14.

15.

16.

The Family of t Distributions

The quantities that make up a t distribution are all numbers that would be calculated in accord with the formula above for a t ratio. You could generate the numbers as follows: Start with a population of scores. Draw from it at random a sample of a certain size. From the sample, find the mean, __17: symbol?__ , and estimate the standard deviation of the population, σ. The estimate is called __18: symbol?__ . Then use this estimate of the standard deviation of the population to estimate the standard deviation of the sampling distribution of means for samples of whatever size you drew; the standard deviation of this sampling distribution is called the __19__ of the mean, and the estimate of its standard error is symbolized $s_{\overline{X}}$. The formula is $s_{\overline{X}} = s/\sqrt{n}$. Finally, for this first sample, calculate the t ratio, using the formula above—with the true value of the population mean in the role of μ_{hyp}.

17.

18.

19.

Now repeat this process a second time, drawing another sample of the same size from the same population. This time, almost certainly, you would get a different value for $\overline{X}$, the __20__ of the sample, and a different value for s, the estimate of the standard __21__ of the population of scores. Thus you would almost certainly get a different value for $s_{\overline{X}}$ and t.

20.

21.

Repeat the process a third time, and a fourth, and continue indefinitely. The resulting collection of t ratios would form one member of the family of t distributions.

What would the distribution look like? It would have a mean of __22__ , and it would have the symmetrical shape of a bell, but it would not have a normal shape. Compared with a normal distribution, a t distribution has __23: more or less?__ area in its tails. See Figure 12.1 on p. 214 for the kind of shape a t distribution possesses.

22.

23.

Sample Size and Degrees of Freedom

The t distribution we just envisioned is only one of a large number (a family) of such distributions. If we change the sample size, we change the nature of the distribution a bit. The mean will always be __24__ , but the shape will change. If the sample size becomes smaller, the shape will be even __25: more or less?__ normal; if sample size becomes larger, the shape will be __26: more or less?__ normal. As sample size approaches infinity, the shape becomes increasingly similar to that of the __27__ .

24.

25.
26.
27.

More generally speaking, the shape of a t distribution depends on a quantity called the degrees of __28__ , which is closely related to sample size. The __29__ of freedom for a given sample are the number of independent pieces of information that the sample provides for purposes of statistical __30: description or inference?__ .

28.
29.

30.

In the present case, we are estimating the standard deviation of the population, __31: symbol?__ , by calculating the quantity called s, and we are then using the estimate s to calculate $s_{\bar{X}}$, an estimate of the __32__ of the mean. The degrees of freedom available from a single sample for computing s and $s_{\bar{X}}$ are one less than the size of the sample. In symbols, $df = $ __33__ .

31.
32.

33.

The table giving critical values of the various t distributions, Appendix G, designates a particular distribution not by the size of the sample involved but by the degrees of freedom, __34: symbol?__ .

34.

The Assumption of Population Normality

Exactly what is it that is not distributed normally in cases of the kind we are considering? It is easy to think that it is the sampling distribution of means whose shape is not normal, whose shape instead follows the distribution of Student's __35__ . This is not so. The quantity whose sampling distribution follows Student's t *is* the t ratio. In fact, t ratios follow Student's distribution exactly only when the sampling distribution of means is itself exactly __36__ . And the sampling distribution of means

35.

36.

will be exactly normal only if the population of raw scores is __37__ . (See   37.
pp. 179-181 to review this matter.)

What if the population is not normal in shape? If sample size is
reasonably large, over 30 or 40, it doesn't matter; taking critical values
from the t table in Appendix G will still yield fairly accurate results even
if the population deviates markedly from normality. If sample size is
small, though, we should examine the data for clues that the population's
shape differs sharply from a unimodal, symmetrical one. If we find such
clues, we should use an alternative to the __38__ test. Some alternatives are   38.
described in Ch. 22 of the text.

Levels of Significance vs. *p*-Values

The procedure for testing a hypothesis, as it is presented in Chs. 11 and
12, is a highly formal one. It requires the investigator to make an explicit
choice of the level of significance, __39: symbol?__ , *before* collecting data. In   39.
practice, some investigators examine their results with no precise level of
__40__ in mind, or if they have one in mind they may not report it. (If they   40.
have one in mind that is unreported, it is usually .05, though sometimes it
is as high as __41__ or as low as __42__ .)   41.
    42.
Investigators who examine their results with no specific level of significance in mind, or who fail to report a specific level, usually do report
what are known as *p*-values. A *p*-value is a measure of how rare the
sample result would be if the null hypothesis, __43: symbol?__ , were   43.
__44: true or false?__ . Specifically, the *p*-value of a sample result is the   44.
probability, if H_0 were true, of observing a sample result as deviant as, or
more deviant (in the direction(s) specified by H_A) than, the result
actually obtained. Knowing the *p*-value, anyone can decide for him- or
herself whether to declare a result statistically significant according to
any given level of significance. That is, anyone can decide for him- or
herself whether to accept or to reject the null hypothesis by applying the
decision criterion known as the level of significance, __45: symbol?__ .   45.
Investigators who report *p*-values seldom say exactly what they are.
Instead they usually report them relative to the landmarks of .05 and .01
and sometimes .10 and .001. If an investigator considers a result to be
significant (that is, if the investigator __46: accepts or rejects?__ the null   46.
hypothesis on the basis of the result), the investigator reports the *p*-value
as less than a landmark (for example, as $p < .05$). If the investigator
considers a result to be nonsignificant (and thus __47__ the null on the basis   47.
of the result), the investigator reports the *p*-value as greater than the
landmark (for example, as p __48__ .10).   48.

The t Test for a Single Mean

To test a null hypothesis (H_0) stating a certain value, μ_{hyp}, for the mean of a population with standard deviation σ, we draw a random sample of some particular size, n, from the population, and we compute its mean, $\overline{X}$. Then, to locate $\overline{X}$ within the random sampling distribution of means of its kind,...

we would like to calculate

but the best we can do is to calculate

$$t = (\overline{X} - \mu_{hyp})/s_{\overline{X}}$$

$$z = (\overline{X} - \mu_{hyp})/\sigma_{\overline{X}}$$

could be evaluated with reference to critical values in

would be

must be evaluated with reference to critical values in

estimates

the Standard Error of the Random Sampling Distribution of Means

the Normal Distribution

Student's Distribution of t

equals

equals

is

$\sigma/\sqrt{n}$

$s/\sqrt{n}$

estimates

is based on

a Family of Distributions, each of which has a

$n - 1$ independent pieces of information (scores)

Bell Shape

are

is thicker in the tails than

becomes more similar

the Degrees of Freedom for the t Ratio

to

with an increase in

are symbolized

df

Types of Errors and the Level of Significance

Whenever a null hypothesis is tested, it is possible to make an error by reaching the wrong conclusion about it. Two types of error are possible. If the null is true but we get significant results and reject it, we make what is called a Type _49_ error. What is the risk of making a Type I error? It is given by the level of significance, _50: symbol?_ . That is, if the null hypothesis is really true, the probability of rejecting it (wrongly) is whatever α value we use for our test of the null. If we set α at .05, the probability of rejecting the null if the null is true is _51_ .

It is important to understand that working at the .05 level of _52_ does not mean that we will make a mistake in hypothesis testing 5% of the time. Choosing $\alpha = .05$ means rather than *in whatever instances the null hypothesis is true* the probability of wrongly _53: accepting or rejecting?_ it is .05.

The other kind of error occurs when the null hypothesis is false. If it is false, we should _54: accept or reject?_ it, and if we accept it, we commit what is called a Type _55_ error.

What can we do to control the risk of making Type I and Type II errors in testing hypotheses? One thing we can do is to choose a good level of significance. If we set the level very high (at .25, say), we run too high a risk of making a Type _56_ error. But if we set it very low (at .0001, say) as a way to reduce the risk of a Type I error, we raise the risk of making a Type II error. The levels of significance in common use, _57_ and _58_ , are compromises. They give us reasonable assurance that we will not reject H_0 when it is really true (that is, they give us good protection against Type _59_ errors). But at the same time, they are not so small that they unnecessarily raise the risk of accepting a false H_0 (which would be a Type _60_ error).

49.
50.

51.
52.

53.

54.
55.

56.

57.
58.

59.

60.

SELF-TEST

The formula for a t ratio is ... A) $(X - \overline{X})/s$. B) $(X - \mu)/\sigma$. C) $(\overline{X} - \mu)/\sigma$. D) $(\overline{X} - \mu)/s_{\overline{X}}$.

1.   A B C D

Compared to the normal distribution, a distribution of t ratios has ... A) a different mean. B) less symmetry. C) more area in the tails. D) less area in the tails.

2.   A B C D

The degrees of freedom available in a sample of size n for computing an estimate of the standard deviation of the parent population are ... A) n. B) $n - 1$. C) $\overline{X}$. D) $\overline{X} - 1$.

3.   A B C D

A distribution of *t* ratios will have a normal shape if ...   A) the sampling distribution of means is normal.   B) sample size is infinite.   C) the *df* are small.   D) the null hypothesis is true.

4.   A B C D

Which depends on (is determined by) the specific outcome that a researcher obtains in a sample?   A) The level of significance for testing a hypothesis about the mean of the population from which the sample came.   B) The *p*-value of the outcome.   C) Both of these.   D) Neither of these.

5.   A B C D

The levels of significance most commonly used are ...   A) .10 and .05.   B) .10 and .01.   C) .05 and .01.   D) .01 and .001.

6.   A B C D

A researcher always uses the .01 level of significance. Which is true for this researcher?   A) One percent of all the researcher's hypothesis tests yield an erroneous conclusion about the null hypothesis.   B) When the null hypothesis is false, the researcher runs a 1% risk of accepting it.   C) The probability that the researcher will make a Type II error is .01.   D) In the long run, the researcher will reject 1% of the true null hypotheses that she or he tests.

7.   A B C D

Raising the level of significance (from .05 to .10, for example), ...   A) raises the risk of rejecting a true null hypothesis.   B) raises the risk of accepting a true null hypothesis.   C) lowers the risk of a Type I error.   D) raises the risk of a Type II error.

8.   A B C D

In a study of a sample of 10 cases, the critical value of *t* is +3.250. If the study had been of 11 cases, but everything else (such as α) had been the same, the critical value would have been ...   A) a little smaller.   B) a great deal smaller.   C) a little larger.   D) a great deal larger.

9.   A B C D

t ratios follow the distribution of Student's *t* exactly only when ...   A) the samples that generated the *t* ratios were selected at random.   B) the population of raw scores has the normal shape.   C) both of these things are true.   D) neither of these things is true.

10.   A B C D

CHAPTER 13

COMPARING MEANS OF TWO POPULATIONS: INDEPENDENT SAMPLES

OBJECTIVES

Chs. 11 and 12 introduced the basic principles and procedures of hypothesis testing and showed you how to test a hypothesis about the mean of a single population. Ch. 13 now presents the test of a hypothesis comparing two populations. In studying the chapter, work to understand the following symbols, terms, concepts, principles, and techniques:

Assumptions Underlying the *t* Test of the Difference between Two Means for Independent Samples
  Normality of the Populations
    The Central Limit Theorem   •   The Role of Sample Size
  Homogeneity of Variance
  Randomization
    Benefits   •   Difficulty in Interpreting Results when Groups were not Formed by Randomization
  Statistical Inferences vs. Non-Statistical Generalizations
    Under the Random Sampling Model   •   Under the Randomization Model

Also in studying Ch. 13, work to learn these skills:

  Selection of the Null Hypothesis for an Investigation Comparing Two Groups of Scores
  Selection of the Alternative Hypothesis for an Investigation Comparing Two Groups of Scores
  Determination of whether the Two Groups are Independent or Dependent
  Execution of a *t* Test Comparing the Means of Two Populations when Samples are Independent, Given Raw Scores
    Computation of the *t* Ratio   •   Calculation of the Degrees of Freedom   •
    Use of Appendix G to Determine the Critical Value(s) of *t*   •   Decision on H_0

REVIEW

Comparing Two Samples

There are many studies in which a researcher compares two groups of observations. In an experiment, for example, a researcher may collect observations of subjects who are treated in some special way (who are in the "experimental condition") and compare them with observations of subjects who were given no special treatment (who are in the "control condition"). In almost all such studies, the researcher must acknowledge that the two groups of observations are each only a sample from a larger   1 . After all, the researcher could have observed more subjects in each condition.

    1.

Now even if the two populations were identical, the researcher must expect some difference between the two samples, just because of the   2   factors involved in the selection of a sample. Finding a difference

    2.

between two samples thus does not guarantee a difference between their parent populations. To draw an inference about the populations from any difference between the two samples, a researcher usually employs inferential statistics to test a hypothesis about the difference between the two _3_ .

3.

The Statistical Hypothesis

If the observations are scores, the researcher will probably use the t-test to compare the mean of one sample with the mean of the other sample. The null hypothesis for this test almost always states that there is no difference between the two _4: sample or population?_ means. In symbols, H_0 says that $\mu_1 - \underline{\ 5\ } = 0.$*

4.

5.

The alternative to the null hypothesis may again be nondirectional or directional. A nondirectional alternative says that the difference between the population _6_ is not zero, allowing both for the possibility that it is greater than zero and for the possibility that it is _7_ zero. This nondirectional, or _8_ -tailed, alternative is appropriate when we wish to discover *either* whether the mean of the first population exceeds the mean of the second population *or* whether the mean of the first is lower than the mean of the second.

6.

7.

8.

A directional alternative hypothesis takes one of _9: how many?_ forms. One form says that $\mu_1 - \mu_2 > 0$ (that is, the first mean is larger than the second). The other form of H_A says that μ_1 _10_ (that is, the first mean is less than the second). A directional hypothesis is appropriate when the researcher is interested in discovering only whether the difference between the first and the second _11: sample or population?_ mean is in a certain "direction."

9.

10.

11.

The Basic Logic of the Test

Having stated a null hypothesis and an alternative to it, a researcher proceeds much as in the test of a null about a single population. The researcher selects a sample from each population and describes the outcome of the sampling in terms of the difference between the two sample means. That is, the researcher calculates ($\overline{X}_1 - \overline{X}_2$). This difference is a number like -5.8 or $+12.1$. To show that we are thinking of

---

*The statistical terminology that calls the hypothesis to be tested a null hypothesis is appropriate here, because the term *null* means zero, and the hypothesis for comparing two populations says that the difference between their means is zero.

$\overline{X}_1 - \overline{X}_2$ as a specific number and not as a subtraction problem, it is helpful to put the expression in parentheses, like this: ($\overline{X}_1 - \overline{X}_2$).

The researcher must then ask whether the obtained difference between the sample means, __12: symbol?__ , is one that would be likely to occur if the null hypothesis were __13: true or false?__ , or whether the obtained difference between the sample means would be rare if the null were true. If the answer is "likely," the researcher should __14__ the null hypothesis; if "rare," the researcher should __15__ the null.

To determine whether the obtained difference between the sample __16__ , ($\overline{X}_1 - \overline{X}_2$), is likely or rare, the researcher will use a criterion called the level of significance, __17: symbol?__ , which is usually .05 or .01. If .05, for example, the researcher regards as a likely difference between the sample means one whose probability of occurrence, according to the null, is __18: greater or less?__ than .05, and as a rare difference between the sample means one whose probability of occurrence, according to the null, is .05 or __19: more or less?__ .

In its basic logic, then, the test of a hypothesis about two population means is identical to the test of a hypothesis about a single population mean.

12.

13.

14.

15.

16.

17.

18.

19.

The Random Sampling Distribution of Differences between Two Means

To determine precisely whether the obtained difference between the sample means, __20: symbol?__ , is one that is likely or one that is rare according to the null hypothesis, the researcher must compare ($\overline{X}_1 - \overline{X}_2$) with the quantities of this kind that would occur if the null were __21: true or false?__ . And the quantities of this kind are the random sampling distribution of differences between __22__ .

You could generate a random __23__ distribution of differences between two means as follows: Start with two populations of scores. Select a sample from each at random. Calculate the mean of each sample, $\overline{X}_1$ and $\overline{X}_2$, and then find the difference between the two, __24: symbol?__ . Now repeat this procedure exactly. Draw from the first population a second sample of whatever size you first drew from that population. Draw from the second population a second sample of whatever size you first drew from *that* population. (The two sample sizes, n_1 and n_2, do not have to be equal.) Calculate the mean of each new sample and then the difference, __25: symbol?__ . Repeat this process forever. The resulting collection of quantities of the kind ($\overline{X}_1 - \overline{X}_2$) would be the random __26__ distribution of differences between two means for the circumstances of your sampling.

20.

21.

22.

23.

24.

25.

26.

In practice, of course, no one could generate such a distribution. But we need not even try, because mathematicians have discovered principles that specify the important characteristics of such a distribution.

1. *The mean of a random sampling distribution of differences between means:* If the null hypothesis is true, so that $\mu_1 - \mu_2$ really equals zero, then the mean of the __27__ distribution of __28__ between means will also be zero. The symbol for the mean of the sampling distribution is __29__ ("mew sub eks bar sub one minus eks bar sub two"). This principle thus says that:

27.

28.

29.

$$\mu_{\bar{X}_1 - \bar{X}_2} = 0 \quad \text{when } H_0 \text{ is true}$$

2. *The standard deviation of a random sampling distribution of differences between means:* This quantity is called the standard __30__ of the difference between means, and its value depends on whether the two samples are independent or dependent. The samples are __31__ when there is no logical way to pair the scores in one sample with the scores in the other. Independence will obtain when the samples are selected at random from their populations, or when a pool of subjects is divided at __32__ into two treatment groups. The description above of the procedure for generating a random sampling distribution of differences between means was for independent samples. (The next chapter treats the case of dependent samples.)

30.

31.

32.

When the samples are independent, the __33__ error of the difference between means is determined by the standard error of the first mean, __34: symbol?__ , and the standard error of the second mean, __35: symbol?__ . The symbol for the standard error of the difference between two means is __36__ ("sigma sub eks bar sub one minus eks bar sub two"). In symbols and in terms of the two standard errors of the mean, the formula for this quantity is:

33.

34.

35.

36.

$$\sigma_{\bar{X}_1 - \bar{X}_2} = 37$$

37.

Since $\sigma_{\bar{X}_1}^2 = \sigma_1^2/n_1$ and $\sigma_{\bar{X}_2}^2 = \sigma_2^2/n_2$, as explained below, this formula tells us two important things about the standard error of the difference between two means. A larger size for either or both samples leads to a __38: larger or smaller?__ standard error. And a greater variability in either or both populations leads to a __39: larger or smaller?__ standard error.

38.

39.

There are complications in using the samples to estimate the standard error of the difference between means. These are noted below.

3. *The shape of a random sampling distribution of differences between means:* If each population is normal, the shape will be exactly __40__ . If

40.

either or both populations is not normal, the shape will only approximate a normal distribution, but with larger samples the shape will be increasingly close to normal, in accord with the __41__ Theorem. If the two sample sizes, __42: symbols?__ , are each at least 20 to 30, the sampling distribution of differences between means will usually be very close to normal in its shape.

41.

42.

Estimating the Standard Error of the Difference between Means

As noted above, the standard error of the difference between means is determined by the standard error of the first mean, $\sigma_{\bar{X}_1}$, and the standard error of the second mean, __43: symbol?__ . The formula is:

43.

$$\sigma_{\bar{X}_1 - \bar{X}_2} = \sqrt{\sigma_{\bar{X}_1}^2 + \sigma_{\bar{X}_2}^2}$$

This formula is in terms of the *squared* standard error of the first mean, $\sigma_{\bar{X}_1}^2$, and the *squared* standard error of the second mean, __44: symbol?__ . You have already learned that the *un*squared standard error of a mean, $\sigma_{\bar{X}}$, equals $\sigma/\sqrt{n}$. If we square both sides of this little equation, we have $\sigma_{\bar{X}}^2 = \sigma^2/n$. Substituting σ^2/n back into the formula immediately above, we get:

44.

$$\sigma_{\bar{X}_1 - \bar{X}_2} = \sqrt{\sigma_1^2/n_1 + \sigma_2^2/n_2}$$

This reworking of the formula helps us understand how to estimate the standard error of the __45__ between means. We assume that the two populations have the same variance; that is, we assume that $\sigma_1^2 = $ __46__ . Now we use our two samples to construct the best estimate we can of this variance. You have already learned that we can best use a sample to estimate the standard deviation of a population by dividing the sum of the squared deviations about the mean of the sample by one less than the sample size. (See p. 195 of the text.) The formula is:

45.

46.

$$s = \sqrt{\Sigma(X - \bar{X})^2/(n - 1)} = \sqrt{\Sigma x^2/(n - 1)}$$

where s is the estimate of the populations's __47__ . If we now square this expression, we will have an estimate of the variance of the population. (Remember that a variance is the square of a standard deviation, so an estimate of a variance is the square of an estimate of a standard deviation.) Using the symbol s^2 for the estimate of the variance of a population, we obtain:

47.

$$s^2 = \Sigma(X - \bar{X})^2/(n - 1) = \Sigma x^2/(n - 1)$$

Let us now express this variance estimate in slightly more general terms. The quantity on top is a sum of squared deviations, and the quantity on the bottom is the degrees of __48__ associated with the sum of squared deviations. (See pp. 212-213 if you wish to review the concept of degrees of freedom.) Thus:

$$s^2 = \Sigma x^2 / (n - 1) = \text{sum of squared deviations/degrees of freedom}$$

and this quantity estimates σ^2, the __49__ of the __50__ .

Returning finally to the problem at hand, which is estimating the variance that we assume is common to the two populations of interest, we can make the best use of our two samples if we combine the sums of squared deviations from their means and divide by their total degrees of freedom in the samples. The combination of the sums of squared deviations is $\Sigma x_1^2 + \Sigma x_2^2$, and the total degrees of freedom is $(n_1 - 1) + (n_2 - 1) = n_1 + n_2 - 2$. Putting the top and the bottom of the fraction together, we get the formula for the pooled variance estimate, symbolized __51__ ("little es squared sub pee"):

$$s_p^2 = 52$$

In the last step of this long line of thought, we substitute s_p^2 for the quantities of which it is an estimate, σ_1^2 and σ_2^2, in the formula cited above for the standard error of the difference between means, $\sigma_{\bar{X}_1 - \bar{X}_2}$. When we substitute the estimate for the real things, we end up with an estimate of $\sigma_{\bar{X}_1 - \bar{X}_2}$; this estimate is symbolized __53__ ("little es sub eks bar sub one minus eks bar sub two"). The whole formula is:

$$s_{\bar{X}_1 - \bar{X}_2} = 54$$

48.

49.

50.

51.

52.

53.

54.

Expressing the Location of ($\bar{X}_1 - \bar{X}_2$) within its Sampling Distribution

In the test of a hypothesis about a single mean, we had to express the location of the obtained mean, $\bar{X}$, within the sampling distribution of those quantities of this kind that would occur if the null hypothesis were __55: true or false?__ . Similarly, in testing a hypothesis about the difference between two population means, we must express the location of the obtained difference between the two sample means, __56: symbol for the difference?__ , within the sampling distribution of those quantities of *this* kind that would obtain if the null hypothesis were true. And as in the case of one mean, we would like to use a z score to do this, but we must content ourselves with an approximation to a z, again called a __57__ ratio.

55.

56.

57.

The formula is:

$$t = 58 \qquad \qquad \text{58.}$$

Concluding the t Test of the Difference between Means

To evaluate our t ratio (or any other), we must compare it with the sampling distribution that it would come from if the null hypothesis were true. This sampling distribution is a member of the family of Student's distribution of t; it is the distribution for the degrees of freedom, _59: symbol?_ , that characterize our t ratio, which are _60_ . (These are the   59.
degrees of freedom specifically for the pooled variance estimate and for   60.
the estimated standard error of the difference between the two means.)

In the sampling distribution of t for the applicable degrees of freedom, we must locate the region or regions of rejection. These are determined by the level of _61_ that we chose and by the nature of our   61.
62: null or alternative? hypothesis—whether it is directional or nondirec-   62.
tional. The procedure is the same as for the case of one mean. We then
reject the null hypothesis if the obtained t ratio falls in the region or
regions of rejection; other we _63_ the null.   63.

Evaluating the Size of the Difference

If the difference between our two sample means permits us to reject the
_64_ hypothesis of no difference between the means of the parent   64.
populations, we are entitled to call the difference between the sample
means statistically significant (given whatever _65_ of significance, α, we   65.
used). But a statistically significant difference is not necessarily a large or
important one. If sample size is very _66_ , even a very small difference   66.
between the two sample means can justify the rejection of the null
hypothesis. It is therefore important to note the size of the difference
between the two sample means.

Sometimes we do not know the variable under study well enough to
tell, just by looking at the difference between the sample means, whether
it is large or small. In this case, we should ask how much of a _67_ the   67.
difference amounts to. We should estimate the standard deviation of the
populations (we have assumed them to be the same), and we can do this
by calculating the square root of the variance estimate: $\sqrt{s_p^2}$. We then
divide the difference between the _68: sample or population?_ means,   68.
69: symbol? , by this quantity. A difference of .2 standard deviation is   69.
small; a difference of .5 standard deviation is moderate; and a difference
of .8 or more is rather large.

The Assumptions of Normality and Homogeneity of Variance

Underlying the procedure described in this chapter for testing a null hypothesis about two means, there are two assumptions. One is that each _70: sample or population?_ of scores is normally distributed. Fortunately, when samples consist of more than 20 to 30 cases, the populations can depart considerably from normality of shape without seriously affecting the validity of the t test.

70.

The second assumption is that the two populations have the same degree of variability. This is the assumption of homogeneity of _71_. Research has shown that the t test will still give valid results, even if this assumption is wrong, unless the two variances are quite different, the two samples are of quite different size, *and* either or both of the samples are small.

71.

Thus small samples can make it hazardous to rely on the t test, though they do not necessarily do so. When samples are small, we should look at the data to see if one or both populations seem to be highly skewed in _72_ and to see if the populations differ drastically in _73_. If either is the case, there are alternatives to the standard t test that we can use.

72.
73.

The Role of Randomization

The t test presented in this chapter is commonly applied to experiments in which a researcher divides a group of subjects into two subgroups via the process of _74_. This process has two benefits.

74.

The first is statistical. Randomization makes it possible to apply the principles that describe natural variation among samples, so we can determine how much difference between two samples can be reasonably attributed just to _75_.

75.

The second benefit of randomization is that it provides experimental control over _76_ factors that might bias a test of the effect of a particular treatment. Suppose, for example, that we are studying some aspect of behavior that is influenced by the subjects' motivation (how hard they try). We thus want the subjects in our two groups to be equal, at least on the average, in the extent of their motivation. If we assign the available subjects to one group or the other at random, it will be _77: likely or unlikely?_ that the more highly motivated subjects will generally end up in one group rather than in the other. Both groups will probably consist of some strongly motivated subjects and some weakly motivated subjects, and the average degree of motivation in the two groups will be about the same. The beauty of randomization is that it affords this type of

76.

77.

experimental control over *all* extraneous factors, and it does so whether or not the researcher knows that the factors exist.

Random assignment of subjects to treatment groups 78: guarantees or does not guarantee? that the groups will be equal in extraneous factors. But randomization does tend to produce equality, and the tendency increases in strength as sample size 79: increases or decreases? .

78.

79.

Generalizing from Samples

Most of the procedures for statistical inference that behavioral scientists employ, including the procedures covered in this chapter, assume that a sample on hand was selected 80: how? from a population of interest. If indeed the sample was chosen in this way, the statistical procedures provide a way of making inferences about the 81 (for example, about its mean). But such generalizations are usually not the only inferences that a researcher wishes to make. A researcher often wants to generalize beyond the original population from which she or he sampled. A generalization of this kind is not statistical; it goes beyond what statistics can show and requires judgment and interpretation based on knowledge of the 82: statistical or substantive? area under study. Inferential procedures of the statistical kind are a necessary first step toward such a generalization, but they are only the first step.

80.

81.

82.

Generalizing in the Case of Randomization

As noted in Section 10.2 on pp. 172-173, the 83: descriptive or inferential? techniques that behavioral scientists employ are usually based on the random sampling model, and the scientists use this model even when they employ randomization rather than random sampling. Applying the random-sampling model to observations generated by randomization is satisfactory, because the random-sampling model will usually yield the same conclusion about a null hypothesis that a model based on 84 would.

83.

84.

There *is* an important limitation to the randomization model, though. Whereas in random sampling the statistical conclusions may properly be generalized to the larger population that was sampled, under the randomization model, the statistical conclusions do not go beyond the particular subjects who were studied. A researcher may still generalize beyond these subjects, and a researcher almost always does, but in such a case the researcher will be making one of those 85: statistical or nonstatistical? generalizations discussed just above. And as a nonstatistical generalization, it must be based on the researcher's knowledge of the substantive area and on his or her good judgment.

85.

Comparing Means of Two Populations via Independent Samples

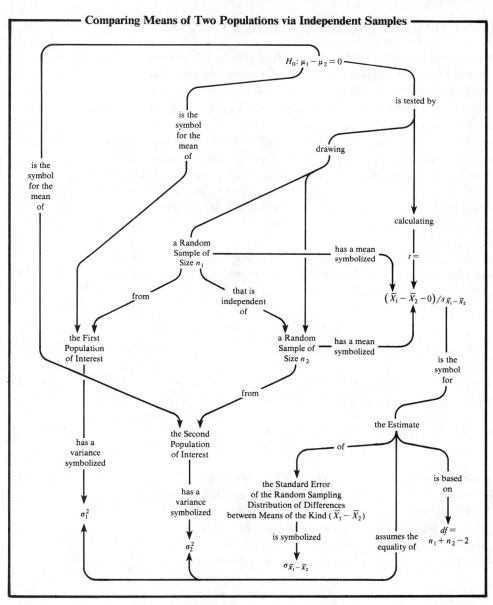

CAUTIONS CONCERNING CONFUSABLE QUANTITIES

Ch. 13 of your text has introduced the symbol $\mu_{\bar{X}_1 - \bar{X}_2}$ ("mu sub eks-bar-sub-one minus eks-bar-sub-two"). It designates the mean of a sampling distribution composed of quantities of the form $(\bar{X}_1 - \bar{X}_2)$, which are each a difference between the mean of a first sample, $\bar{X}_1$, and the mean of a second sample, $\bar{X}_2$. (The subscript to the symbol μ names these quantities.)

If you are careless in how you write this complex symbol, you may produce something that looks like $\mu_{\bar{X}_1} - \bar{X}_2$. This is a bizarre expression designating the difference between $\mu_{\bar{X}_1}$, the mean of a sampling distribution of sample means called $\bar{X}_1$, and the mean of a sample called X_2. Even if you become a professional statistician, you will probably have no occasion to deal with such an expression. Be sure you write $\mu_{\bar{X}_1 - \bar{X}_2}$ neatly enough so you don't think it refers to anything like that bit of nonsense.

A similar problem can arise in careless writing of the symbol for the standard error of a sampling distribution of differences between means, $\sigma_{\bar{X}_1 - \bar{X}_2}$. Don't mistake it for $\sigma_{\bar{X}_1} - \bar{X}_2$. There is no such animal.

SELF-TEST

Which is the usual null hypothesis for a comparison of two means?
A) $\mu_1 = \mu_2$.   B) $\bar{X}_1 = \bar{X}_2$.   C) $\mu_1 \neq \mu_2$.   D) $\bar{X}_1 \neq \bar{X}_2$.

1.   A   B   C   D

Which could be an alternative hypothesis for a comparison of two means?   A) $\mu_1 = \mu_2$.   B) $\bar{X}_1 > \bar{X}_2$.   C) $\mu_1 < \mu_2$.   D) $\bar{X}_1 < \bar{X}_2$.

2.   A   B   C   D

A researcher is doing a two-tailed test comparing two means. The researcher finds that $(\bar{X}_1 - \bar{X}_2)$ is $+17.5$, and that if the null hypothesis is true, the probability of getting a value for $(\bar{X}_1 - \bar{X}_2)$ that is $+17.5$ or greater is less than the level of significance the researcher had selected. The researcher should ...   A) accept the null hypothesis.   B) reject the null hypothesis.   C) change the level of significance.   D) switch to a one-tailed test.

3.   A   B   C   D

Two populations of scores have the same mean. The mean of the random sampling distribution of differences between means of samples drawn from these populations is ...   A) impossible to predict accurately. B) the same as whatever is the value of the populations' means. C) zero, but only if the populations are both normal in shape. D) zero under all circumstances.

4.   A   B   C   D

The standard error of the random sampling distribution of differences between means is ...   A) a kind of standard deviation.   B) determined by the means of the two populations from which the samples came. C) determined by the shapes of the two populations from which the samples came.   D) all three of these things.

5.   A   B   C   D

The differences that make up a random sampling distribution of differences between means are ...   A) differences between the means of two samples.   B) differences between the means of two populations. C) differences between the mean of a sample and the mean of a population.   D) a mixture of these three kinds of quantity.

6.   A   B   C   D

Under what circumstances will the random sampling distribution of differences between means be exactly normal?   A) When both samples have fewer than 25 cases.   B) When either or both samples have more than 25 cases. C) When both populations are exactly normal. D) When either or both populations are exactly normal.

7.   A   B   C   D

The general formula for an estimate of a variance is sum of squared deviations divided by ...   A) population size.   B) sample size.   C) level of significance.   D) degrees of freedom.

8.   A   B   C   D

The t test for comparing two means of independent samples assumes that ...   A) both populations are normally distributed.   B) the two populations have the same variability.   C) both of these things are true. D) neither of these things is true.

9.   A   B   C   D

Which procedure makes possible the wider generalization of the statistical kind?   A) Random sampling.   B) Randomization.   C) Both procedures limit generalization of the statistical kind only to the subjects actually studied.   D) The procedures permit equally broad generalizations beyond the subjects actually studied.

10.   A   B   C   D

CHAPTER 14

COMPARING MEANS OF TWO POPULATIONS: DEPENDENT SAMPLES

OBJECTIVES

Ch. 13 presented the t test for the comparison of two means when the samples are independent of one another. Ch. 14 now tells you how to proceed when the samples are dependent. In studying this chapter, work to understand the following symbols, terms, concepts, principles, and techniques:

Also in studying this chapter, work to learn these new skills (which supplement those taught in the previous chapter):

Execution of the t Test Comparing Dependent Means, Given Raw Scores
Computation of the Estimate of the Standard Error of the Difference between
Means for Dependent Samples   •   Computation of the Degrees of Freedom
for this Estimate
Execution of a t Test Comparing Dependent Means, Using Difference Scores
Statement of H_0 and H_A •   Calculation of Difference Scores   •   Calcu-
lation of the t Ratio

REVIEW

Research Designs that Generate Dependent Samples

In general, two samples are dependent if the observations in one are __1__   1.
to the observations in another—that is, if any given score in one sample
bears a special relationship to a particular score in the other sample.
Specifically, two samples are dependent if there is a logical way to pair
each score in one sample with a score in the other sample. Dependence
can arise in two ways:

1. Each subject is observed under 2: only one or both? of the condi-   2.
tions of a study with two conditions. Each subject will thus contribute
 3: one score or two scores? to the data. There will be a first score for one   3.
condition and a second score for the other condition. The hypothetical
study of dieters described in this chapter generated this kind of dependence. Here there is only one group of subjects but two samples of
(paired) scores.

2. Dependence 4: can also or cannot? occur if the subjects in one   4.
group are totally different from the subjects in the other group. If a
researcher begins an investigation by pairing the subjects—begins, that
is, by identifying pairs of subjects who are equal or approximately so on
some characteristic—and then tests one member of each pair under one
condition and the other member of the pair under the other condition,
the researcher will collect two samples of scores that are dependent. Here
there are two groups of subjects and two samples of (paired) scores.

The Standard Error of the Difference between Dependent Means

The t test comparing two means for the case in which the samples are
dependent is highly similar to the t test comparing two means for the case
in which the samples are _5_. There are only three differences.   5.

A first difference occurs in the formula for the standard __6__ of the difference between means. In the previous chapter (on p. 233), you learned that the standard error of the difference between independent means can be estimated from the samples by forming a pooled estimate of the variance we assume to be common to the two populations. We use the pooled variance estimate, __7: symbol?__ , in the following formula:

$$s_{\bar{X}_1 - \bar{X}_2} = 8$$

In the analogous formula for dependent samples, there is a term that includes the coefficient of __9__ between the scores in the first sample and the scores in the second sample. This coefficient, __10: symbol?__ , will usually be positive, because pairs of scores in which one score is high will typically be pairs in which the other score is __11: low or high?__ , and pairs in which one score is low will typically be pairs in which the other score is __12: low or high?__ .

The term that includes the correlation coefficient, r, is __13: added to or subtracted from?__ the other quantities in the formula for estimating the standard error of the difference between dependent means. As a result, when samples are dependent, the standard error of the difference between means will normally be __14: larger or smaller?__ than it is when samples are independent.

The formula for the case of dependent samples is:

$$s_{\bar{X}_1 - \bar{X}_2} = 15$$

In this formula, $s_{\bar{X}_1 - \bar{X}_2}$ is the __16__ of the standard error of the difference between means, s_1^2 is the estimate of the __17__ of the scores in the first population, s_2^2 is the estimate of the variance of the scores in the second population, r_{12} ("are sub one two") is the correlation between the scores in the first sample and those in the second sample, s_1 is the estimate of the __18__ of the first population, __19__ is the estimate of the standard deviation of the scores of the second population, and n is the number of __20: scores or pairs of scores?__ .

There is a statistical advantage to using dependent rather than independent samples, and it lies in the reduction of __21: symbol?__ . The smaller this standard error, the greater the likelihood that any difference between the two sample means will accurately reflect whatever difference there may be between the two population means.

The amount of the reduction in the standard error depends on the size of __22: symbol?__ . The stronger the correlation (assuming it to be positive), the smaller the value of __23: symbol?__ , and the greater the advantage to using dependent samples.

6.

7.

8.

9.

10.

11.

12.

13.

14.

15.

16.
17.

18.
19.
20.

21.

22.
23.

Degrees of Freedom in the Case of Dependent Samples

The second difference between the t test for independent samples and its counterpart for dependent samples lies in the degrees of freedom for the t ratio. For independent samples, the degrees of freedom, __24: symbol?__ , are $n_1 + n_2 - 2$, where n_1 is the number of scores in the first sample, and n_2 is the number in the second sample. For dependent samples, $df = $ __25__ , where n is the number of *pairs* of scores.

24.

25.

An Alternative Approach to the Problem of Two Dependent Means

The third difference between the t test for independent samples and that for dependent ones lies in the availability of an alternative method of computing the t ratio when the samples are __26: independent or dependent?__ . If we use the alternative method, instead of dealing explicitly with the two distributions of scores and calculating the correlation coefficient for the paired values, we focus on a single distribution. This distribution consists of __27__ between paired X_1 and X_2 scores. That is, for each pair, we compute a quantity, symbolized D, that is the difference between the first and the second score of the pair: $D = (X_1 - X_2)$.

26.

27.

We now apply to the one distribution of difference scores the t test of a hypothesis about the mean of a single population. In this case, our interest lies in the mean of the population of difference scores, of course. What hypothesis should we test? The null for comparing the two populations from which we started in our sampling says that __28: symbols?__ . If this null were true, then the mean of the population of differences between the paired values would also be __29__ . So, in the alternative approach to the comparison of dependent means, the null hypothesis says that $\mu_D = 0$.

28.

29.

We must also state a(n) __30__ hypothesis, as usual. Again, the hypothesis will depend on its counterpart for the test comparing the two populations that were our initial concern. If the alternative for the two populations says that $\mu_1 - \mu_2 \neq 0$, then the alternative for the one-distribution test should say that $\mu_D \neq 0$; if $\mu_1 - \mu_2 > 0$, then $\mu_D > 0$; if $\mu_1 - \mu_2 < 0$, then μ_D __31__ 0. (These translations assume that each difference score, __32: symbol?__ , is calculated as $X_1 - X_2$ rather than as $X_2 - X_1$.)

30.

31.

32.

We then proceed *exactly* as we would for the t test of a hypothesis about the mean of any one population. The only departure from the

procedure that you studied in Ch. 11 is that we are now calling the scores
D instead of __33__.   33.

The t ratio in our alternative approach takes the form:

$$t = \left(\bar{D} - \mu_{D(\text{hyp})} \right) / 34$$ 34.

Here $\bar{D}$ is the mean of the __35: sample or population?__ of difference scores;   35.
$\mu_{D(\text{hyp})}$ ("mew sub dee hype") is the mean of the population of difference
scores stated by the __36__ hypothesis, which is 0; and $s_{\bar{D}}$ is the estimate of   36.
the standard error of the random sampling distribution of means.

Look back at Table 11.1 on p. 197 for the original example of a
one-sample test, and you will see that the present test is identical except
for the name of the scores (__37__ instead of X).   37.

If we do the arithmetic correctly, we will find that $s_{\bar{D}}$ turns out to be
the same as __38: symbol?__, and the t ratio turns out to be __39: similar to or__   38.
__the same as?__ the t calculated from the two-distribution approach. The df 39.
are still __40: formula?__.   40.

Using the alternative approach reduces the computational work for t
substantially, but it yields __41: more or less?__ information about the sam-   41.
ples. In most research we will want to know (and we will be obliged to
report) the mean and the standard deviation of each of the two original
samples, and often we will also want to know __42__. If we use the   42.
alternative approach to the t test, we will have to go back to the original
data and compute these other quantities. Thus we may find that the total
amount of work is about the same in the two approaches.

Some Pros and Cons of Dependent Samples

As noted above, there is a statistical advantage in using dependent
samples rather than independent ones, and it lies in the reduction of the
__43__ of the difference between __44__. A smaller standard error makes it   43.
more likely that we will properly reject H_0: $\mu_1 - \mu_2 = 0$ in cases where   44.
H_0 is indeed false.

A research design employing dependent samples can, however, cause a
variety of difficulties in both the statistics and the interpretation of the
results. The best procedure is to match subjects into pairs that are similar
on some attribute related to the one under study, and then use the
process of __45__ to assign one member of each pair to each of the two   45.
conditions of the study. The randomization should be done separately for
each pair. Such a procedure combines the virtue of matching subjects and
the virtue of randomizing assignment of subject to condition.

Difficulties occur, however, if the design of the study departs from this ideal combination of matching and 46 . If we observe the same person under both treatment conditions, rather than using 47 pairs of subjects, it is possible that the exposure to the condition the subject encounters first will systematically change the subject in some way, altering his or her performance under the second condition. If so, a(n) 48 effect is going on, and we should try to control for it by dividing the subjects into two groups, administering the two conditions to half the subjects in one order and to the other half of the subjects in the other order. (The revised design, then, would have four conditions.)

46.

47.

48.

Difficulties also occur if we depart from the ideal combination of matching and randomization by matching without randomization. Interpretation of any difference between the two samples of scores is obscured by the possibilities that we did either too little matching or too much.

These difficulties should not, however, cause us to reject out of hand a research design that will generate dependent samples of observations. Such a design can be very useful. But we should be aware that problems may arise, and we should try to prevent or solve them before we expend time and effort in collecting observations.

ANNALS OF EGREGIOUS* EXAMPLES

A businessman who ran a marketing-research firm once consulted the author of this *Guide* for advice on the analysis of some data he'd collected. He had conducted two studies testing consumer reaction to several varieties of frozen food. In both studies, his subjects were shoppers who were approached in public places such as malls and parking lots. The subjects were asked to taste one or more varieties of a food (which had been cooked, of course) and to report a judgment of "bad," "poor," "fair," "good," or "excellent." The fellow had translated these judgments into the numbers 1 through 5.

What is of interest here is a subtle difference between the two studies. To simplify things a bit, let us suppose that only two varieties of a food were compared in each study. In the first one, there were a total of 200 subjects. Half of them had tasted one variety and half the other variety; a given subject had tasted only one, so there were a total of 200 scores.

You now know that it would be foolish to conclude that any difference between the two samples must indicate a difference between their parent populations. You know that inferential statistics is needed to draw a proper conclusion about the

---

*Egregious ("eh-GREE-juss"): conspicuously bad.

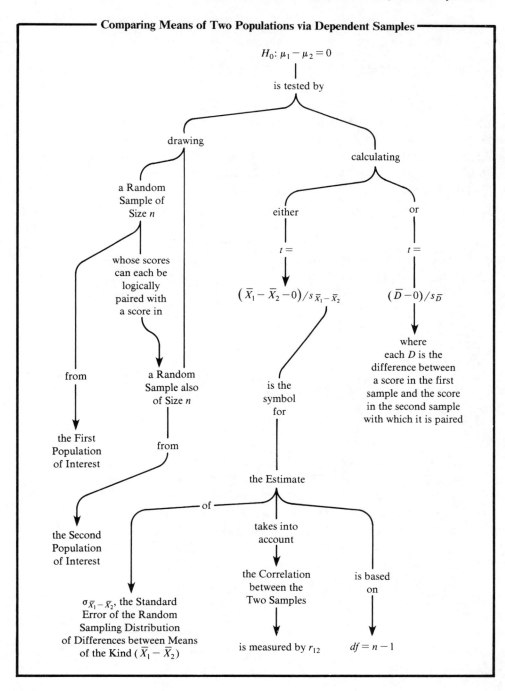

Comparing Means of Two Populations via Dependent Samples

$H_0: \mu_1 - \mu_2 = 0$

is tested by

drawing

calculating

a Random Sample of Size n

either

or

whose scores can each be logically paired with a score in

$t =$

$t =$

$(\bar{X}_1 - \bar{X}_2 - 0)/s_{\bar{X}_1 - \bar{X}_2}$

$(\bar{D} - 0)/s_{\bar{D}}$

from

a Random Sample also of Size n

is the symbol for

where each D is the difference between a score in the first sample and the score in the second sample with which it is paired

the First Population of Interest

from

the Second Population of Interest

the Estimate

of

takes into account

is based on

$\sigma_{\bar{X}_1 - \bar{X}_2}$, the Standard Error of the Random Sampling Distribution of Differences between Means of the Kind $(\bar{X}_1 - \bar{X}_2)$

the Correlation between the Two Samples

is measured by r_{12}

$df = n - 1$

populations, and you know how to analyze these data so you could draw such a conclusion. Say how you would analyze the scores by answering the questions below.

1. What statistics would you calculate to describe the data?

2. What inferential procedure would you apply? Say whether you would test a hypothesis about a single population mean or about two population means, and if the latter, whether the sample means are independent or dependent. State your null hypothesis and your alternative hypothesis, choosing between a one-tailed and a two-tailed test. List the calculations you would have to do.

In the second study, now, there were only 100 subjects, but each subject had tasted *two* varieties of a food and judged both of them, so each subject contributed two scores to the data. There were again a total of 200 scores. You also know how to analyze data of this kind. Again outline how you would do it by answering the questions below.

3. What statistics would you calculate to describe the data?

4. What inferential procedure would you apply? Spell out all the details as for Question 2 above.

Now we come to the reason why this example is egregious. In his first study, the businessman had collected 200 judgments, 100 recorded on one page of a notebook for one variety of a food, and the other 100 recorded on a second page for a second variety. The fellow had cast each sample of 100 scores into a frequency distribution, producing two tables looking like these (the frequencies are hypothetical):

Judgments of Variety A		Judgments of Variety B	
Score	f	Score	f
5	23	5	17
4	55	4	40
3	12	3	23
2	6	2	15
1	4	1	5
	$\Sigma f = 100$		$\Sigma f = 100$

The mean and the other descriptive statistics required for each sample were easy to calculate from these tables. (Your text covers these techniques in Problem 15 on p. 83*.) Things were fine for the first study.

But this is exactly how the businessman had preserved his data for the second study too. Two tables of this sort were all that he had to go on.

5. Something is wrong here. What is it?

The data from the second study, then, could not be properly analyzed. The businessman could hardly believe it; the difference between the procedures he'd followed in the two studies seemed so slight.

Moral: *Plan the statistical techniques you'll use on your observations **before** you collect and record those observations.*

The businessman confessed to me that he had already written his report on the foods for the company that was considering marketing them. In the report he had simply asserted that the variety with the highest mean rating in each study was *signficantly* higher than the others tested in that study—but he didn't really know this to be so, not even for the first study. *A course in statistics helps preserve one's honesty.*

SELF-TEST

In studying the effectiveness of a new method of psychotherapy, a researcher plans to compare an experimental condition in which the therapy was employed with a control condition in which it was not. The researcher haphazardly gathers a group of people suffering from shyness, and he pretests each person for the exact degree of shyness. He then finds the two who are most shy, and he flips a coin to determine which of them will receive therapy and which will go on a waiting list for therapy. He next identifies the two who are next most shy, and again he flips a coin to determine which of the two goes into which condition of his study. He continues in this fashion until he has identified the two who are least shy and has again flipped the coin to determine which of the two will receive therapy and which will wait for it. Which of the following procedures has this researcher used?   A) Matching.   B) Randomization.   C) Both of these.   D) Neither of these.

1.   A   B   C   D

*For the kind of ungrouped scores the businessman was working with, a raw score, X, would serve in the role of the midpoint of a class interval, X_m, in the formulas cited in the problem.

Which of the procedures on the following list did the researcher of Question 1 employ? A) Random sampling. B) Testing the same subject in both conditions of the study. C) Both of these. D) Neither of these.

2.   A   B   C   D

The researcher of Question 1 will retest each subject for shyness after a certain interval of time; during this interval the subjects in the experimental condition will have undergone the new kind of psychotherapy, while those in the control condition will have received no special help. The retesting will generate two samples of scores. The samples will be ... A) independent. B) dependent. C) unrelated to each other. D) impossible to arrange into logically formed pairs of scores.

3.   A   B   C   D

If a first sample of scores comes from one group of subjects and a second sample of scores comes from an entirely separate group of subjects, ... A) the two samples must be independent. B) the two samples must be dependent. C) the two samples may be independent or dependent; there is not enough information here to tell. D) it is necessary to do a t test that deals with two distributions of scores.

4.   A   B   C   D

A researcher plans to collect two samples, each of 25 scores. If the researcher's design leads the samples to be dependent rather than independent, the standard error of the difference will probably be_____, and the degrees of freedom will definitely be_____   A) larger ... larger. B) larger ... smaller. C) smaller ... larger. D) smaller ... smaller.

5.   A   B   C   D

The stronger the correlation between the scores in one sample and the scores in the other sample (assuming a positive relationship), ...   A) the larger the standard error of the difference between means. B) the smaller the standard error of the difference between means. C) the larger the degrees of freedom. D) the smaller the degrees of freedom.

6.   A   B   C   D

Suppose we have two samples, each of 10 scores, and they are dependent samples. If we use the "alternative approach" to the comparison of their populations, the approach that focuses on a single distribution, rather than the approach that works with two distributions, we must ... A) calculate a difference score for each pair of raw scores. B) use a different level of significance. C) expect the degrees of freedom for the t ratio to be different. D) expect the value of the t ratio to be different.

7.   A   B   C   D

Which can be a null hypothesis for a comparison of two populations when we have dependent samples? A) $\mu_1 - \mu_2 = 0$. B) $\mu_D = 0$. C) Either of these. D) Neither of these.

8.   A   B   C   D

Using samples that are dependent rather than independent is advantageous statistically, because (if everything else is equal), when samples are dependent, ...   A) the standard error of the difference between means will be larger.   B) the degrees of freedom will be larger.   C) the likelihood of rejecting the null hypothesis if it is false will be larger.   D) the difference between the population means specified by the null hypothesis will be larger.

9.   A   B   C   D

In which of these research designs is it easiest to properly interpret the results?   A) The researcher observes each subject twice, once in one condition and once in the other condition.   B) The researcher forms matched pairs of subjects and assigns the two members of each pair to one condition or the other at random.   C) On a certain first characteristic, the researcher matches a person who has already been subjected to one condition with a person who has already been subjected to another condition. The researcher forms a number of such pairs. The researcher then compares these two intact groups of paired subjects for a second characteristic.   D) Interpretation of the results of these three designs is equally easy.

10.   A   B   C   D

CHAPTER 15

STATISTICAL POWER ANALYSIS

OBJECTIVES

Ch. 15 deepens your understanding of hypothesis testing by introducing the concept of statistical power and presenting some principles of statistical power that you can put to good use either in doing research yourself or in interpreting the results of other people's research. In studying Ch. 15, work to understand the following symbols, terms, concepts, principles, and techniques:

The Power of a Test
Effect Size
   d
Factors Affecting Power
  Effect Size
  Sample Size
    Implications for Choice of Sample Size
  Level of Significance
  Use of Dependent Samples
  One-Tailed vs. Two-Tailed Test
Power Curves
  Effect on the Curves of Figure 15.3 of Changing:
    to a One-Tailed Test   •   to Dependent Samples   •   to the .01 Level
    of Significance
Significance vs. Importance
  Risk when Samples are too Small   •   Risk when Samples are too Large
Selection of Sample Size
  Minimum Important d
Relation between Power and Type II Error
Relation between Power and Type I Error

Also in studying this chapter, work to learn this skill:

Determination of the Power of a t Test Comparing Two Means, via Figure 15.3

153

REVIEW

Statistical Power

Every test of a null hypothesis possesses an important characteristic called its power. The power of a statistical test is the probability, given that the null hypothesis of the test is 1: true or false? , of obtaining sample results that lead to the 2: acceptance or rejection? of the null hypothesis. To put it less formally, a powerful test is one that offers us a high probability of claiming that there is a difference between what the null hypothesis says and what reality is—when there is in fact 3: such or no such? a difference.

1.

2.

3.

A Measure of Effect Size: *d*

To understand the factors that affect the power of a statistical test, we need a way to measure the discrepancy between the state of affairs envisioned by a null hypothesis and the true state of affairs. In this chapter, discussion focuses on the test of the difference between two means (though the principles introduced here are quite general), so the null hypothesis states that 4: symbols? . If the null is false, then the difference between the two population means, $\mu_1 - \mu_2$, is not zero, and to provide a frame of reference for this difference, we can make good use of the standard deviation of the populations.

4.

We make the assumption of homogeneity of 5 ; that is, we assume that the two populations have the same variance. In symbols, we assume that $\sigma_1^2 = $ 6: symbol? , and we use the symbol σ^2 (without a subscript) for whatever this number is. The standard deviation is the square root of the 7 , so the standard deviation common to the two populations will have the symbol 8 . Using this symbol now, we measure the difference between the two population means as 9 . This quantity is symbolized *d*, and it is called the 10 .

5.

6.

7.
8.
9.
10.

If μ_1 is less than μ_2, *d* will be negative, but in this case we should ignore the minus sign. Our measure of effect size is the value of the number without regard for whether it is positive or negative.

Should the standard deviations of the populations be not identical but only similar, we can use the mean of the two standard deviations as an approximation.

Note that it is the standard deviation of the scores that we are using here, and not a standard deviation of a sampling distribution, which would be called a standard 11 .

11.

Factors Affecting Power: Effect Size

We can now consider some factors that affect the power of the *t* test comparing two means. One factor is intuitively obvious: the bigger the difference between the two population means, the more likely the *t* test is to "recognize" this difference when we draw a random sample from each population and use the test to compare the sample means. In more formal terms, using the effect size __12: symbol?__ as a measure of the difference between the population means, the relevant principle is this: the larger the effect size, *d*, the __13: greater or lower?__ the power of a test of H_0: __14: symbols?__ against the nondirectional alternative H_A: __15: symbols?__.

    The same principle holds for a one-tailed test, but with the qualification that the true difference, $\mu_1 - \mu_2$, must be __16: opposite to or in?__ the direction specified by H_A.

12.
13.
14.
15.

16.

Factors Affecting Power: Sample Size

The factor affecting power just noted, effect size, is beyond the control of an investigator. But there is a second factor that an investigator can do something about: the size of his or her samples. Sample size influences power by altering the size of the standard error of the difference between means, __17: symbol?__. When samples are larger, the standard error of the difference is __18: larger or smaller?__, and our estimate of the standard error, __19: symbol for estimate?__, will thus generally be __20: larger or smaller?__. Because this estimate is the denominator of the *t* ratio, when samples are larger, we expect the *t* ratio to be __21: larger or smaller?__ (either positively or negatively). With a larger *t* ratio, in turn, we are more likely to reject H_0. Thus for any given effect size other than __22__, the larger the sample size, the greater the power of the test.

17.
18.
19.
20.
21.

22.

    We have now seen that big differences between two population means and large samples from those populations make for a __23: more or less?__ powerful test, while small differences and small samples make for a __24: more or less?__ powerful one. Consequently, if we're looking for a difference that might be small, we will need to choose samples that are __25: small or large?__ enough. But if we would be interested in discovering a difference only if it is large, it may be possible to do so with small samples.

23.

24.

25.

Factors Affecting Power: Level of Significance

A third factor influencing the power of a test is the level of significance, __26: symbol?__, which like sample size is under the researcher's control.

26.

The larger the value of α, the larger the regions of   27: acceptance   27.
or rejection? , and thus the greater the power of the test. Inversely, the
smaller the value of α, the less the power. In light of this principle,
researchers seldom use a very low level of significance, such as .001.
Though it would reduce the risk of making a Type 28 error, it would   28.
greatly diminish the power of the test.

Factors Affecting Power: Use of Dependent Samples

A fourth factor influencing the power of a test is again one under the
investigator's control. This is the use of dependent rather than 29   29.
samples. As Section 14.3 on p. 251 pointed out, when we use dependent
samples, we will almost always enjoy a smaller standard error of the 30   30.
between means. The amount of reduction will depend on the degree of
dependence, which is measured by the coefficient of 31 between the   31.
two samples. And with a smaller standard error, if the null hypothesis is
false, the t ratio will tend to be 32: larger or smaller? , leading to a greater   32.
chance of correctly rejecting the null hypothesis. Thus the use of dependent samples will usually 33: increase or decrease? the power of a test of   33.
$H_0: \mu_1 - \mu_2 = 0$. The amount of increase depends on the degree of 34 .   34.

Factors Affecting Power: One-Tailed vs. Two-Tailed Tests

A fifth factor influencing power, and yet again one under the control of a
researcher, is the choice between a directional and a nondirectional
 35: null or alternative? hypothesis. If we get the direction of the dif-   35.
ference between the means correct in our alternative hypothesis, then
choosing the one-tailed variety will 36: increase or decrease? the power of   36.
our test.

On the basis of this principle, some people have argued for greater use
of one-tailed tests. The authors of your text recommend, however, that
the choice of alternative hypothesis should flow from the logic of the
investigation.

Power Curves

Tables and charts are available to tell us what the power of a statistical
test would be for various possible effect sizes. Figure 15.3 on p. 274
shows such a chart for the kind of test that this chapter focuses on, the t
test comparing means of two samples. The chart assumes that the

populations have the same degree of variability and that the two samples are of the same size.

The curves in this figure are for _37: one-tailed or two-tailed?_ tests, for   37.
independent samples, and for the _38_ level of significance. For a   38.
one-tailed test where the direction of the difference between the means
specified by H_A is correct, the power will be _39: higher or lower?_ . The   39.
increase will be small to moderate, depending on n and on d—at most
about .15 higher than the value obtained from the chart. For dependent
samples, power will also be higher, depending in great part on the degree
of the _40_ . The stronger the relationship between the two samples, the   40.
higher the power of the test. But if we use the .01 rather than the .05 level
of significance, power will be somewhat _41: higher or lower?_ .   41.

Significance vs. Importance

Statistical jargon can be misleading. As you have learned, calling the
result of a test "significant" implies that the researcher discovered
something important, but the term *significant* here means only that the
researcher _42: accepted or rejected?_ a null hypothesis. We must be espe-   42.
cially wary when samples are large, because large samples may give
 43: significant or nonsignificant? results even when the population dif-   43.
ference is too small to be important.

On the other hand, samples that are too small tend to give
 44: significant or nonsignificant? results, even when the population dif-   44.
ference *is* large enough to be important.

Selection of Sample Size

The principles just enunciated tell us that we should select samples that
are large enough to give us a good probability of detecting an important
difference between the two populations of interest. But the samples
should not be too large, or we will waste time and effort or discover a
statistically significant difference between the samples when the difference between the populations is minor. How then do we select samples
of the appropriate size?

The first step is to decide what is the smallest discrepancy between the
hypothesized value and the value that we wish to be reasonably certain of
discovering. In other words, we must determine what is the minimum
value of _45: symbol?_ that is large enough to be important to us.   45.

Statistical Power and Types of Error

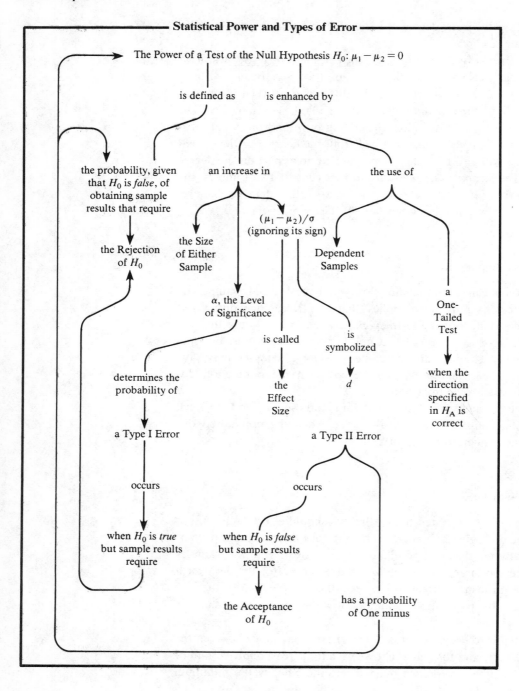

The Power of a Test of the Null Hypothesis H_0: $\mu_1 - \mu_2 = 0$

is defined as

is enhanced by

the probability, given that H_0 is *false*, of obtaining sample results that require

an increase in

the use of

$(\mu_1 - \mu_2)/\sigma$ (ignoring its sign)

the Rejection of H_0

the Size of Either Sample

Dependent Samples

a One-Tailed Test

α, the Level of Significance

is called

is symbolized

when the direction specified in H_A is correct

determines the probability of

the Effect Size

d

a Type I Error

a Type II Error

occurs

occurs

when H_0 is *true* but sample results require

when H_0 is *false* but sample results require

has a probability of One minus

the Acceptance of H_0

We then decide what we want for the minumum __46__ of detecting this effect.

46.

Putting together the value of d and the probability, we use Figure 15.1 to determine the sample size.

Power and Type I Error

As you have learned, in testing a null hypothesis, we can make either of two errors. If H_0 is true and we __47__ it, we commit a Type I error. If H_0 is false and we __48__ it, we commit a Type II error. The power of a statistical test is closely related both to the risk of a Type I error and to the risk of a Type II error.

47.

48.

Consider first the risk of a Type I error. As Ch. 12 informed you, α, the __49__, is the probability of making a Type I error when the null hypothesis is true. In this chapter now, you saw that α affects the power of a test: the larger the value of α, the __50: higher or lower?__ the power. Thus the less we guard against a Type I error (by switching from .05 to .10, say), the __51: higher or lower?__ the power of our test. And the more we guard against a Type I error (as by switching from .05 to .01), the __52: higher or lower?__ the power.

49.

50.

51.

52.

Power and Type II Error

The relationship between power and the risk of a Type II error is even simpler. Power is the probability of __53: accepting or rejecting?__ a null hypothesis when it is false, and the risk of a Type II error is the probability of __54: accepting or rejecting?__ the null when it is false. The only two possibilities are acceptance and rejection, and so their probabilities must sum to unity (one). Thus Pr(Type II error) = __55__.

53.

54.

55.

SELF-TEST

The power of a statistical test is the probability of obtaining sample results that lead us to ...   A) accept H_0 when it is true.   B) accept H_0 when it is false.   C) reject H_0 when it is true.   D) reject H_0 when it is false.

1.   A B C D

Suppose $\mu_1 = 55$, $\mu_2 = 35$, $\sigma_1 = \sigma_2 = 10$, $\sigma_{\bar{X}_1} = 2.0$, $\sigma_{\bar{X}_2} = 5.0$, and $\sigma_{\bar{X}_1 - \bar{X}_2} = 5.4$. What is d in this case?   A) 2.0   B) 4.0   C) 10.0   D) None of these.

2.   A B C D

Other things equal, the power of a statistical test increases if d is ... and sample sizes are ... A) larger ... larger. B) larger ... smaller. C) smaller ... larger. D) smaller ... smaller.

3. A B C D

Sample size influences the power of a test because sample size affects ... A) $\mu_1 - \mu_2$. B) $\bar{X}_1 - \bar{X}_2$. C) σ. D) $\sigma_{\bar{X}_1 - \bar{X}_2}$.

4. A B C D

If we use dependent samples rather than independent ones in a t test comparing two means, the power will normally be _____, and the amount of the change in power will depend on the size of _____ A) greater ... $\mu_1 - \mu_2$. B) greater ... $\bar{X}_1 - \bar{X}_2$. C) greater ... r. D) smaller ... r.

5. A B C D

If everything else stays the same when we switch from a two-tailed test to a one-tailed test, the power of the test _____ if the direction of the difference between the population means specified in H_A is _____ A) increases ... correct. B) increases ... wrong. C) decreases ... correct. D) stays the same ... wrong.

6. A B C D

Suppose there is a difference between the mean of a first population and the mean of a second, but it is a trivial difference. If we don't want to discover such a small difference, what should we do? A) Keep the risk of a Type II error low. B) Use a very powerful statistical test. C) Use relatively small samples. D) Do all of the above.

7. A B C D

If the difference between two population means is trivial, there is a certain kind of error that we shouldn't mind making when we test a hypothesis about the difference between these means. What is it? A) Setting the α level too high. B) Using a test with too much power. C) Making a Type I error. D) Making a Type II error.

8. A B C D

Suppose the power of a statistical test is known to be .83. Which of the following could we calculate from this fact? A) The risk of a Type I error. B) The risk of a Type II error. C) The level of significance. D) The p-value.

9. A B C D

Which of these factors affecting the power of a test is *not* under the investigator's control? A) The level of significance. B) Whether the samples are independent or dependent. C) The size of the samples. D) The size of the effect.

10. A B C D

CHAPTER 16

ESTIMATION

OBJECTIVES

Hypothesis testing, the subject of the last four chapters, is one of two basic methods of statistical inference. The other is estimation, and it is the subject of this chapter. In studying the chapter, work to understand the following symbols, terms, concepts, principles, and techniques:

Differences between Hypothesis Testing and Estimation
Point Estimation
    Disadvantage
Interval Estimation
    Confidence Interval • Confidence Limits • Level of Confidence (Confidence Coefficient)
Interval Estimation of μ
    t_c (e.g., t_{95}, t_{99})
Interval Estimation of $\mu_1 - \mu_2$
Effect of the Confidence Coefficient on the Width of the Interval
Effect of Sample Size on the Width of the Interval
Relationship between an Interval Estimate and a Hypothesis Test
    A Confidence Interval Contains Values that Would Lead to Acceptance of the Null Hypothesis
    The Assumptions underlying Interval Estimation are those for the Comparable Hypothesis Test
        Assumptions for Estimation of μ
        Assumptions for Estimation of $\mu_1 - \mu_2$
            When Samples are Independent • When Samples are Dependent
Advantages of Interval Estimation

Also in studying this chapter, work to learn these skills:

Construction of an Interval Estimate of μ, Given a Sample
Construction of an Interval Estimate of $\mu_1 - \mu_2$, Given:
    Independent Samples • Dependent Samples

REVIEW

The Nature of Estimation

In testing a 1: null or alternative? hypothesis, we ask whether or not a *specific* condition holds in the population or populations of interest to us. We ask, for example, whether the mean of a population, 2: symbol? , is exactly equal to 126. We ask whether the difference between the mean of a first population and the mean of a second is exactly zero. Characteristics of populations such as these are called 3: statistics or parameters? . In hypothesis testing, then, we ask whether the parameter of interest takes a certain value.

In estimation, on the other hand, we have no specific number in mind for the parameter of interest, and we do not test any hypothesis about the 4: statistic or parameter? . Instead, we ask questions such as "What *is* the value of μ?" "What *is* the difference between the population means?" That is, we ask a direct question about the parameter of interest.

Any problem that we can approach through hypothesis testing we can also approach through 5 . And there are problems that lend themselves to estimation but not to hypothesis testing. Estimation is thus a very widely useful technique.

Point Estimation vs. Interval Estimation

An estimate may take two forms: a point estimate or a(n) 6 estimate. A point estimate is simply a single value derived from a sample. If the mean of a random sample from a certain population of IQ scores is 108.4, we can make a 7 estimate of the mean of that population by simply asserting that it is also 108.4. Characteristics of samples such as the mean are called 8: statistics or parameters? , and the technique of point estimation is a simple extrapolation from a statistic to the corresponding parameter.

But we should make a point estimate only with great reluctance, because we know that the 9 factors involved in random sampling will certainly cause such a number to be in error—and we have not tried to determine by how much.

The other kind of estimate, the 10 estimate, is much superior, because this kind is a range of values (an interval) within which we can be reasonably confident that the population parameter lies. How confident can we be that the parameter is within such an interval? That's up to

1.

2.

3.

4.

5.

6.

7.

8.

9.

10.

us to decide. If we choose to establish an interval that merits a great deal of confidence, it will have to be relatively 11: wide or narrow? . If we want the interval to be more narrow and thus more precise, it will merit 12: more or less? confidence.

11.

12.

An interval estimate, like a point estimate, must derive from a sample drawn from any population of interest. But the calculation of an interval estimate is more complex than the simple extrapolation by which we make a point estimate.

Interval Estimates of μ

Suppose we wish to establish an interval estimate for the mean of a certain population, and we want to have 95% confidence that this interval will include the mean. Exactly what would it mean to have 95% confidence in such an interval? The figure 95 would indicate this: if we repeatedly draw random samples from the population of interest, and if we follow a certain rule for constructing an interval estimate on the basis of each sample, then in the long run, 13 % of the estimates so constructed will in fact cover the parameter we want to know, μ. In general, a c% confidence interval is one so constructed that c% of such intervals include the parameter we are after.

13.

How could we calculate a 95% confidence interval for a population mean? We start our reasoning with a basic property of any distribution of values that has a normal shape: 95% of the values are no farther away from the mean than 14 standard deviations. Now we extend this principle to a random sampling distribution of means: if the sampling distribution is normal, then 15 % of the sample means are no farther away from the mean of these means, which is 16 , than 1.96 standard deviations. The standard deviation here is called the 17 of the mean, and it is symbolized 18 .

14.

15.
16.
17.
18.

Now if 95% of sample means are no farther away from μ than $1.96\sigma_{\bar{X}}$, then it is equally true that for 95% of the sample means, μ is no farther away than $1.96\sigma_{\bar{X}}$. Suppose then that for each possible sample from the population of interest, we assert that μ lies somewhere within the range $\bar{X} \pm 1.96\sigma_{\bar{X}}$. For 95% of the sample means, this statement would be correct. In drawing samples at random from the population of interest, then, the probability is .95 that an interval estimate constructed according to the rule $\bar{X} \pm 1.96\sigma_{\bar{X}}$ would include μ.

In practice, of course, we cannot use this rule, because we do not know the standard deviation, 19: symbol? , of the population, and thus we

19.

cannot know the standard error of the mean, <u>20: symbol?</u> , which equals   20.
$\sigma/\sqrt{n}$. We will have to use the sample estimate $s_{\overline{X}} = $ <u>21</u> in place of   21.
$\sigma_{\overline{X}} = \sigma/\sqrt{n}$.

Furthermore, to make allowance for the error introduced by this
procedure, we will have to work with Student's distribution of <u>22</u> rather   22.
than the normal distribution of z scores.

The rule for a 95% confidence interval for μ thus becomes:

$$\overline{X} \pm t_{.95}s_{\overline{X}}$$

In this rule, $t_{.95}$ is that value of t such that the middle <u>23</u> % of the t 23.
distribution for the applicable degrees of freedom lies between $-t$ and
$+t$. That is, $t_{.95}$ is the t beyond which there lies .05 of the area under the
curve in <u>24: either tail or both tails combined?</u> . The applicable degrees of   24.
freedom are <u>25</u> .   25.

If we want our interval to permit 99% confidence, the only change in
the rule is to select the value of <u>26</u> such that 99% of the t distribution   26.
for the applicable df will be included within the limits $-t$ to $+t$. This will
be the value of t beyond which <u>27</u> % of the area falls in the two tails   27.
combined.

Interval Estimates of $\mu_1 - \mu_2$

The logic underlying interval estimation of $\mu_1 - \mu_2$ is basically
<u>28: the same as or different from?</u> that underlying interval estimation of a   28.
single population mean. The rule for estimating the difference between
the population means is this:

$$29 \pm t_c s_{\overline{X}_1 - \overline{X}_2}$$ 29.

In this rule, t_c ("tee sub see") is the value of t such that the middle c%
(usually 95% or 99%) of the t distribution for the applicable degrees of
freedom lies between $-t$ and $+t$. If the samples are independent,
$df = $ <u>30</u> . If the samples are dependent, $df = $ <u>31</u> , where n is the number   30.
of *pairs* of observations.   31.

Interval Width and Sample Size

You have learned that the choice of a confidence coefficient (the percentage figure) influences the width of the resulting interval. If we choose
to construct a 99% confidence interval rather than a 95% interval, we
must pay for the increase in confidence by accepting a <u>32: wider or</u>   32.
<u>more narrow?</u> interval.

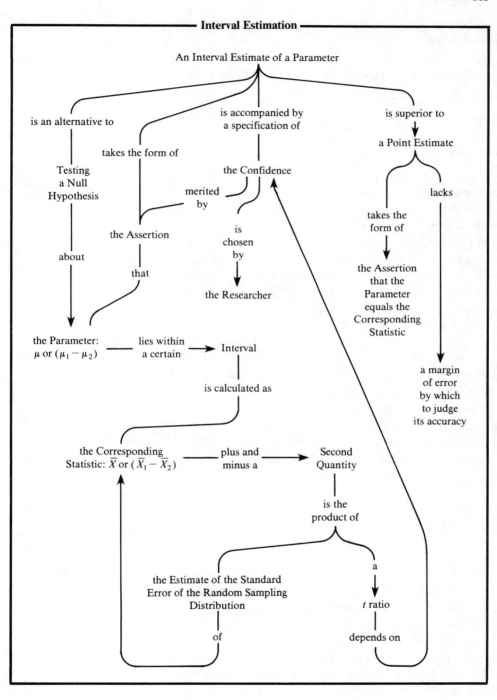

There is something we can do to control the width of an interval so that for any given level of confidence we can estimate the population value within a desired margin of error: we can adjust the __33__ of our sample or samples. If we increase sample size, while other things stay the same, we will get a __34: wider or more narrow?__ interval.

33.

34.

Interval Estimation and Hypothesis Testing

Interval estimation and hypothesis testing are two sides of the same coin. Consider, for example, a 95% confidence interval for a population mean or a difference between two population means. A 95% interval will contain all values for the parameter of interest that, had they been specified in a null hypothesis, would have led to acceptance of the null at the __35__ level of significance in a two-tailed test of the null. Similarly, any hypothesized value for a parameter that falls within a 99% confidence interval would, in a two-tailed test at the __36__ level of significance, have been __37: accepted or rejected?__ .

The close connection between hypothesis testing and interval __38__ extends to the underlying statistical assumptions. Underlying an interval estimate of μ is the same assumption that underlies a test of a null hypothesis concerning μ, the assumption that the population is distributed __39__ . Underlying an interval estimate of $\mu_1 - \mu_2$ are the same assumptions that underlie a test of a null hypothesis concerning $\mu_1 - \mu_2$, the assumptions that both populations are distributed normally and—if the samples are independent—that they have the same __40__ . As with hypothesis testing, the assumptions become important in practice only when samples are __41: large or small?__ .

35.

36.

37.

38.

39.

40.

41.

SELF-TEST

A candidate for the governorship of your state hires you to conduct a poll of the public to determine the percentage of voters who recognize her name. You draw a random sample of the voters and find that 18.3% of the sample say they have heard of the candidate and can identify her as such. You tell the candidate that the percentage of voters across the state who recognize her name is 18.3%. Which technique of statistical inference did you use?   A) Hypothesis testing.   B) Point estimation. C) Interval estimation.   D) None of these.

1.   A   B   C   D

Suppose the admissions committee at your college wants to know the
mean score on the verbal part of the Scholastic Aptitude Test (SAT) for
all people applying for admission to the college. If they cannot study the
whole population of applicants and must rely on a sample, which
technique of statistical inference would they be best advised to use?
A) Hypothesis testing.   B) Point estimation.   C) Interval estimation.
D) Intuition.

2.   A   B   C   D

An interval estimate of a population mean μ ...   A) must always be
accompanied by a statement of the degree of confidence that μ falls
within the interval.   B) must always be accompanied by a statement of
the level of significance for the interval.   C) is calculated from a study
of the entire population.   D) is less valuable than a point estimate of μ.

3.   A   B   C   D

The symbol $t_{.95}$ designates a value of t with which of these properties?
A) 95% of the t values for the applicable df lie above $+t_{.95}$.   B) 95% of
the t values for the applicable df lie below $-t_{.95}$.   C) 47.5% of the t
values for the applicable df lie above $+t_{.95}$ for the applicable df, and
47.5% of the t values for the applicable df lie below $-t_{.95}$.   D) 95% of
the t values for the applicable df lie between $-t_{.95}$ and $+t_{.95}$.

4.   A   B   C   D

The rule for constructing a confidence interval for $\mu_1 - \mu_2$ is ...
A) $\bar{X} \pm t_c s_{\bar{X}}$.   B) $\bar{X} \pm t_c s_{\bar{X}_1 - \bar{X}_2}$.   C) $\bar{X}_1 - \bar{X}_2 \pm t_c s_{\bar{X}}$.   D) $\bar{X}_1 - \bar{X}_2 \pm t_c s_{\bar{X}_1 - \bar{X}_2}$.

5.   A   B   C   D

The df for a confidence interval for the difference between two means,
when the samples are independent, is ...   A) $n_1 + n_2$.   B) $(n_1 - 1) + (n_2 - 1)$.   C) $n - 1$.   D) $n - 2$.

6.   A   B   C   D

If we switch from a 95% confidence interval to a 99% interval, but
everything else stays the same, the interval will ...   A) become more
narrow.   B) stay the same, but the degrees of freedom will change.
C) stay the same, but the level of significance will change.   D) become
wider.

7.   A   B   C   D

If we switch from a small sample to a large one, but everything else stays
the same, a confidence interval constructed on the basis of the sample
will ...   A) become more narrow.   B) stay the same, but the degrees of
freedom will change.   C) stay the same, but the level of significance will
change.   D) become wider.

8.   A   B   C   D

On the basis of a sample, a researcher finds that the 95% confidence
interval for a certain population mean is 88 to 92. If we use the same
sample to conduct a two-tailed test of a null hypothesis about the mean

9.   A   B   C   D

of that population, which of the following hypotheses would we accept?
A) $\mu = 93$.   B) $\mu = 90$.   C) $\mu = 87$.   D) It depends on the level of significance for the test.

If we use the sample described in Question 9, which of the following null hypotheses would we *reject*?   A) $\mu = 94$.   B) $\mu = 93$.   C) $\mu = 87$. D) It depends on the level of significance.

10.   A   B   C   D

CHAPTER 17

INFERENCE ABOUT PEARSON CORRELATION COEFFICIENTS

OBJECTIVES

Ch. 17 shows you how to apply techniques of inference to the Pearson correlation coefficient, r. In studying this chapter, work to understand the following symbols, terms, concepts, principles, and techniques:

The Symbol for the Correlation Coefficient for a Population: ρ (Rho)
Testing a Hypothesis about the Correlation Coefficient for a Population
    The Null Hypothesis
    The Alternative Hypothesis
        Nondirectional   •   Directional
    Critical Value(s) of r
        Degrees of Freedom for r
    Conversion of r to a t Ratio
    Significance vs. Importance
        Factors Affecting the Power of the Test
            Magnitude of ρ •   Sample Size
        Judging the Importance of a Value of r
            For a Large Sample   •   For a Small Sample
Interval Estimation of the Correlation Coefficient for a Population
    The Effect of Sample Size

Also in studying this chapter, work to learn these skills:

Execution of a Test of the Significance of a Value of r, Given the Sample Size
    Selection of the Null Hypothesis
    Selection of the Alternative Hypothesis
    Determination of the Degrees of Freedom
    Determination of the Critical Value(s) of r Using Appendix H
    Via a t test of H_0: $\rho = 0$
        Conversion of r and n to a t Ratio
Construction of a 95% Confidence Interval for ρ, Given r and the Sample Size, via Figure 17.2

REVIEW

Overview

When we calculate the Pearson correlation coefficient for a sample of cases, we are calculating a statistic—that is, a characteristic of a   1: sample or population? . If we drew several samples from the same population and computed the correlation coefficient for each, we would find that this statistic, like others, varies from sample to sample. The variation is due to the chance factors associated with random   2 . To learn the "true" value of the correlation coefficient, which is the   3: statistic or parameter?   characterizing the population, we can use either of the two basic techniques of statistical inference, hypothesis testing or   4 . These techniques as they apply to Pearson correlation coefficients are the subject of this chapter.

1.

2.

3.

4.

Testing the Hypothesis that $\rho = 0$

The question most frequently asked in problems of association is whether there is *any* relationship at all between two variables. That is, in the population of interest, does the correlation equal   5 . The symbol for the Pearson correlation coefficient characterizing a population is   6   ("ro"), which is the lower-case Greek letter rho. The null hypothesis for a test about ρ thus states that   7 . The alternative hypothesis may be nondirectional or directional. A nondirectional alternative states that   8 , and as usual it gives rise to a two-tailed test. A directional alternative states either that $\rho > 0$, or that ρ 9 , and it gives rise to a one-tailed test.

Appendix H of the text is a table listing critical values of r. The table gives values for both the .05 and .01 levels of   10   and for both two-tailed and one-tailed tests. To use the table, we need to know that the degrees of freedom,   11: symbol? , for r are   12 , where n is the number of *pairs* of scores in the sample.

For a two-tailed test, we should reject the   13   hypothesis if the value of r in the sample is equal to or greater than the critical value, whether the obtained value is positive or negative. For a one-tailed test, we should reject the null hypothesis if the obtained value in the sample is equal to or greater than the   14   value *and* is in the direction stated in the   15   hypothesis.

As an alternative to using Appendix H, we can calculate a t statistic using a formula based on the values of r and   16 . The t statistic follows Student's distribution with   17   degrees of freedom, where n is again the number of pairs of scores in the sample.

5.

6.

7.

8.

9.

10.

11.

12.

13.

14.

15.

16.

17.

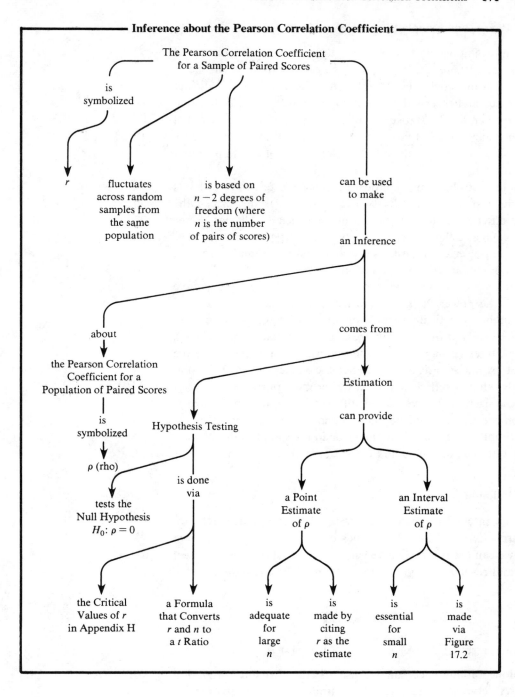

Significance, Power, and Sample Size

Once again, it is important to note that the term *significant* can be misleading in statistical usage. What does a researcher mean when she or he calls a correlation coefficient significant? The expression indicates only that the researcher tested the null hypothesis H_0: $\rho =$ _18_ and   18.
rejected it. In other words, statistical significance here means only that the researcher believes that _19: symbol?_ does not fall precisely at zero.   19.
Whether the value is different enough from zero to be of any importance is a separate matter.

The probability of obtaining significant results in a test of r, which is the _20_ of the test, is affected by two factors. One is the size of the   20.
correlation coefficient in the _21: sample or population?_ . The larger the   21.
value of _22: symbol?_ , the higher the probability of rejecting the null—   22.
that is, the more powerful the test. The second factor is sample size. The larger the sample, the more _23_ the test of H_0: $\rho = 0$.   23.

As a consequence of the fact that the size of ρ and the size of a sample both influence the power of the test, when a sample is very large, the value of r can be statistically significant even though the correlation coefficient is trivially different from zero in the population. When a sample is very large, we should judge the importance of a finding by the size of r, not by its significance. The sample value will give a point estimate of _24_ that is sufficiently accurate for most purposes.   24.

When sample size is small, we should still be concerned with the size of the correlation coefficient in the population and not just with whether the value in the sample is _25_ . But here the value in the sample will not yield   25.
a good point estimate, so we should construct a(n) _26_ estimate of ρ.   26.

Interval Estimation of ρ

To construct an interval estimate of ρ for the 95% confidence level, we can use Figure 17.2 on p. 302. The values of r shown in the figure are positive, but we can use it for negative values too. We should merely treat r as positive and reverse the signs of the _27_ given by the figure.   27.

SELF-TEST

Which of these is a statistic?   A) μ B) ρ C) r D) σ 1.   A   B   C   D

Which technique of inference can we use for the Pearson correlation coefficient?   A) Hypothesis testing.   B) Estimation.   C) Both of these.   D) Neither of these.   2.   A   B   C   D

Which is the question of inference that is asked most frequently in problems of association?   A) Is there any association between the two variables in the sample?   B) Is there any association between the variables in the population?   C) What is the value of ρ?   D) Is the value of r in this sample significantly different from the value of r in this other sample?

3.   A B C D

The null hypothesis for a test of the correlation coefficient states that ...
A) $\rho = 0$.   B) $\rho \neq 0$.   C) $r = 0$.   D) $r \neq 0$.

4.   A B C D

What does the alternative hypothesis for a test of a correlation coefficient state?   A) $\rho \neq 0$.   B) $\rho < 0$.   C) $\rho > 0$.   D) It depends on what the researcher chooses.

5.   A B C D

What are the degrees of freedom for testing a value of r for its statistical significance?   A) n B) $n - 1$ C) $n - 2$ D) It depends on whether the population is distributed normally.

6.   A B C D

The larger the value of ρ, the larger the ...   A) probability of obtaining a significant value of r.   B) probability of accepting the null hypothesis. C) level of significance for the test.   D) degree of confidence in the interval estimate.

7.   A B C D

The larger the size of the sample, the larger the ...   A) power of the test of H_0: $\rho = 0$.   B) probability of obtaining a nonsignificant value of r. C) width of the confidence interval for ρ.   D) α level.

8.   A B C D

To judge the importance of a correlation coefficient obtained from a sample, we should ...   A) take into account the size of the sample. B) construct an estimate of the population value.   C) do both of these things.   D) do neither of these things.

9.   A B C D

For a two-tailed test of the null hypothesis concerning ρ, Appendix H says that the critical value of r is .444 when the df are 18 and the level of significance is .05. If the obtained value of r for 18 df is $-.54$ and we are running a two-tailed test at the .05 level, what should we do?   A) Accept the alternative hypothesis.   B) Accept the null hypothesis.   C) Reject the alternative hypothesis.   D) Reject the null hypothesis.

10.   A B C D

CHAPTER 18

ANALYSIS OF VARIANCE: THE ONE-WAY DESIGN

OBJECTIVES

Ch. 13 presented an inferential technique for comparing the means of two independent samples: the t-test. Ch. 18 now describes an extension of this technique suitable for investigations in which there are more than two samples. In studying the chapter, work to understand the following symbols, terms, concepts, principles, and techniques:

The Type of Study to which the One-Way Analysis of Variance is Applicable
    The Liabilities of Multiple t-Tests   •   The One-Way Analysis of Variance as
    an Extension of the t-Test for Independent Samples
The Null Hypothesis
The Alternative Hypothesis
The Logic of One-Way Analysis of Variance
    Within-Groups Variation
        As a Reflection of Inherent Variation
    Among-Groups Variation
        As a Reflection of Inherent Variation plus any Differential Treatment Effect
    The Relative Size of Among-Groups to Within-Groups Variation as an Indicator of Whether H_0 is False
The Population Variance (σ^2)
    The Variance as Inherent Variation
    The Assumption of Homogeneity of Variance: $\sigma_D^2 = \sigma_E^2 = \sigma_F^2 = ...$
    Estimation of a Population Variance: $s^2 =$ (Sum of Squared Deviations)/(Degrees of Freedom) = (Sum of Squares: SS)/df
The Partition of the Total Sum of Squares
    The Grand Mean: $\overline{\overline{X}}$
    The Total Sum of Squares: $SS_T = \Sigma(X - \overline{\overline{X}})^2$
    The Within-Groups Sum of Squares: $SS_W = \Sigma(X - \overline{X})^2$
    The Among-Groups Sum of Squares: $SS_A = \Sigma(\overline{X} - \overline{\overline{X}})^2$
    $SS_W + SS_A = SS_T$

The Partition of the Total Degrees of Freedom
  The Total Degrees of Freedom: $df_T = n_T - 1$
  The Within-Groups Degrees of Freedom: $df_W = n_T - k$
  The Among-Groups Degrees of Freedom: $df_A = k - 1$
  $df_W + df_A = df_T$
The Variance Estimates
  The Within-Groups Variance Estimate: $s_W^2 = SS_A / df_W$
    As an Estimate of σ^2 •   The Raw-Score Formula
  The Among-Groups Variance Estimate: $s_A^2 = SS_A / df_A$
    As an Estimate of $\sigma^2 +$ Any Differential Treatment Effect   •   The Raw-
    Score Formula
The F Ratio: $F = s_A^2 / s_W^2$
  The Distribution of F when H_0 is True
    As a Family Depending on df_A and df_W
    The Region of Rejection Lies Entirely in the Upper Tail
      Appendix I as a Source of Critical Values
Conclusions about the Null Hypothesis
The Summary Table for an Analysis of Variance
The Assumptions Underlying the F-Test:
  Independence of the Samples   •   Random Sampling   •   Homogeneity
  of Variance   •   Normality of the Populations   •   The Effect of Sample
  Size
Other Tests
  Planned Comparisons
  Post Hoc Comparisons
    Tukey's HSD Test
      HSD   • q

Also in studying Ch. 18, work to learn these skills:

Execution of an F-Test for a One-Way Design, Given Raw Scores
  Statement of H_0 and H_A
  Computation of:
    SS_W and df_W • SS_A and df_A • SS_T and df_T • s_W^2 • s_A^2 • F
  Use of Appendix I to Determine the Critical Value of F
  Decision on H_0
Execution of Tukey's HSD Test, Given the Results of a One-Way Analysis of
Variance
    Use of Appendix J to Determine q •   Computation of the Critical HSD
      •   Determination of which Pairs of Means are Significantly Different

REVIEW

The t-test of the hypothesis of no difference between two population means is very helpful in research, but there is a major limitation to its use: it can compare only two sample means at a time. How, then, shall we analyze the data of the many studies that examine the effect of three or more treatments and thus collect three or more samples of scores, for each of which there is a mean? (An example of such a study would be a comparison of a low dose of a drug with a high dose, using a control condition in which no drug is administered.)

We could press the t-test into service, comparing the various sample means two at a time, but there would be four liabilities to this procedure. A) There would have to be many t-tests. B) In any one comparison, the t-test would employ only the information in the two samples being compared, ignoring information from the other samples that could make the test more sensitive to any difference between the means of the parent populations from which the samples under comparison came. (This sensitivity is called the __1__ of the test.) C) With so many tests, there would be an increased likelihood of committing a Type I error—an increased likelihood, that is, of obtaining one or more __2: significant or nonsignificant?__ differences between sample means when no true difference (no difference between the population means) exists. D) When the various t-tests were completed, there would still be no single, direct answer to the question of whether the treatments, considered as a whole, made a difference in whatever was under study.

Introduced in Ch. 18 is the analysis of variance, a class of inferential techniques that provide an alternative method applicable to two or more means and that suffer from none of the problems of multiple t-tests. The method explained in this chapter is called the __3__-way analysis of variance. It is suitable when the various treatments differ in only one way, as in the example offered above, where the only difference among the treatments lies in the dosage of the drug. (The next chapter presents a type of analysis of variance suitable for studies in which treatments differ in two ways.)

The one-way analysis of variance requires that the samples of scores be __4: dependent or independent?__, and it is in effect an extension of the __5__-test of the difference between two __6: dependent or independent?__ means.

In a study to which the one-way analysis of variance is applicable, the subjects are ideally assigned __7: how?__ to two or more different treatments. The various treatments are symbolized with the letters __8__, and

1.

2.

3.

4.
5.
6.

7.

8.

the number of treatments is symbolized __9__. The mean of the scores   9.
collected from subjects who received Treatment D is symbolized $\overline{X}_D$; the
mean of the scores collected from subjects who received Treatment E is
symbolized $\overline{X}_E$; and so on. Each collection of scores is best regarded as a
sample from a population, and the population means are symbolized
__10__.   10.

The Hypotheses

The null hypothesis in the analysis of variance is a direct extension of the
null for the *t*-test comparing just two means. That null says that both
populations of interest have the same __11__, and the null for the analysis   11.
of variance says that all populations of interest have the same mean. In
symbols, H_0 says that __12__. This null hypothesis would be true if the   12.
various treatments all have the same effect.

The alternative hypothesis states simply that the null hypothesis is
false. It does not say exactly how, and there are many ways in which the
null could be false. For example, just one mean could differ from the
others, and that one could be the mean of Population D, the mean of
Population E, or any of the others. Or all populations could have a
different mean. The distinction between a directional and a nondirectional alternative hypothesis   __13: no longer makes or continues to make?__   13.
sense.

The General Logic of the One-Way Analysis of Variance

Suppose that the several treatments all have the same effect, so the
several populations all have the same mean, and the null hypothesis is
true. What would random samples, one from each population, probably
look like in this case? Figure 18.1 on p. 307 provides an illustration.
Within each sample, the scores vary somewhat about their own mean.
(The scores within Sample D vary about their mean __14: symbol?__ ; the   14.
scores within Sample E vary about __15: symbol?__ ; and the scores within   15.
Sample F vary about __16: symbol?__ .) Variability of this type within a   16.
given sample is called __17__ variation.   17.

How does within-groups variation arise? Because all the subjects
within a give group (sample) received the same treatment (the same
dosage of the drug, for example), the within-groups variation has just one
source, and that is __18__ variation, which is the variability among individ-   18.
uals given identical treatment. (Inherent variation occurs because the
individuals differ in their genetic make-up and in their past history.)

There is a second kind of variability illustrated in the figure on p. 307, and that is the variability among the means of the three samples. Even though the three treatments all had the same effect in this example, so the three population means all take the same value, as the null hypothesis states, the three sample means, $\overline{X}_D$, $\overline{X}_E$, and $\overline{X}_F$, vary somewhat among themselves. This kind of variability is called __19__ variation.

19.

Why does among-groups variation occur? Like within-groups variation, it is due to inherent variation. The greater the inherent variation, the __20__ the variability we will usually find among the sample means.

20.

On the basis of the within-groups variation, we can construct an estimate of the inherent variation characterizing the populations. And we can do the same on the basis of the among-groups variation too. If the null hypothesis is true, as in Figure 18.1, both estimates should yield about the same value.

But now suppose the several treatments do *not* all have the same effect, so the population means are *not* all equal, and the null hypothesis is false. This state of affairs is illustrated in Figure 18.2 on p. 308. Now what we would expect to find in random samples from the populations of interest? As the figure shows, we would again discover variation across the scores in any given sample. In fact, this __21__ -groups variation __22: is or is not?__ affected by whether the null hypothesis is true or false. Because all subjects within a given group (all subjects who contribute a score to a given sample) receive the same treatment, within-groups variation *always* reflects only __23__ variation.

21.
22.

23.

But things are not so simple for the among-groups variation. A random sample will usually have a mean in the vicinity of its parent population's mean, as p. 308 shows, so if the treatments do not all have the same effect and the population means are thus not all equal, the sample means will tend to be unequal too. And this cause for variability among the sample means will augment the variability due to inherent variation. Thus among-groups variation reflects inherent variation *plus* any differential __24__ effect.

24.

If the null hypothesis is false, then, what will happen when we construct those two estimates of the inherent variation in the populations? The estimate derived from the __25__ -groups variation is likely to be larger than the estimate derived from the __26__ -groups variation.

25.
26.

The logic underlying the analysis of variance should now be clear. From the samples on hand, we construct two estimates of the __27__ variation in the populations. The first estimate is based on the variability of individual scores about their own sample means, which is called the __28__ variation. The second is based on the variability among the sample

27.

28.

means, which is called the __29__ variation. If the two estimates are about   29.
the same size, we will have discovered just the state of affairs that is most
likely when the null hypothesis true, and we will __30: accept or reject?__ the   30.
null as possibly correct. But if the estimate of the __31__ variation derived   31.
from the among-groups variation is larger than the estimate derived from
the __32__ variation, and so much so that sampling variation cannot   32.
reasonably account for it, we will __33__ the null.   33.

Estimating σ^2

To proceed with the analysis of variance, we make the assumption that
the inherent variation is the same under all treatments, so that all
populations have the same variance. This is the assumption of __34__ of   34.
variance. Thus we assume that the variance of Population D,   35.
__35: symbol?__ , equals the variance of Population E, __36: symbol?__ , and the   36.
variance of Population F, __37: symbol?__ , and so on for however many   37.
populations are involved in the study (one for each treatment).

The quantity that (we assume) is the variance common to all the
populations is the quantity that we will estimate in two ways from our
samples. To represent it, we use the symbol for the variance of a
population with no subscript on the symbol: __38__ .   38.

To construct both estimates of this variance, we will follow the strategy
introduced on p. 233 of Ch. 13. From the data on hand, we will calculate
the deviations of some numbers about their mean, square those deviations, sum the squared deviations, and divide the sum by the appropriate
degrees of freedom. The estimate will have the usual symbol, which is
__39__ . All that is new here is the terminology: the sum of the squared   39.
deviations is called a __40__ and is symbolized __41__ . Thus the general   40.
formula for a variance estimate is this: 41.

$$s^2 = (\text{sum of squared deviations}) / (\text{degrees of freedom}) = \text{42: symbols?}$$ 42.

The Within-Groups Quantities

The estimate of σ^2 derived from the within-groups variation is called the
within-groups variance estimate and is symbolized __43__ . The sum of   43.
squares (that is, the sum of squared deviations) that contributes to this
estimate is called the within-groups sum of squares, symbolized __44__ . The   44.
degrees of freedom for the estimate is called the __45__ -groups degrees of   45.
freedom, symbolized __46__ .   46.

To calculate the within-groups sum of squares, <u>47: symbol?</u> , we must find for every score in the samples the amount by which it deviates from the mean *of its own sample*. The deviation is calculated in the usual way, as the score minus the mean, or in symbols as <u>48</u> . We must then square each such deviation, and when we have all the squares, we must sum them. In symbols

$$SS_W = 49$$

In using this formula, it is essential to understand that the sample mean, $\overline{X}$, that we subtract from a given score does not remain the same throughout the whole calculation. If we start with Sample D, at first we subtract <u>50: symbol?</u> from each raw score X (and we go on to square each of the resulting deviations). But when we get to Sample E, now we must subtract from each raw score in that sample the mean of Sample E, whose symbol is <u>51</u> . The left-hand portion of Figure 18.1 on p. 312 makes the pattern of the calculations clear.

Thus the expression for SS_W could be written as follows: $\Sigma(X - \overline{X}_D)^2$ for Sample E + $\Sigma(X - \overline{X}_E)^2$ for Sample F + $\Sigma(X - \overline{X}_F)^2$ for Sample F (and so on for additional samples).

With the expression for SS_W in this form, it is easy to see how the degrees of freedom are calculated. Each sample has degrees of freedom equal to one fewer than the <u>52</u> . The total degrees of freedom for the within-groups variance estimate is thus $(n_D - 1) + (n_E - 1) + (n_F - 1)$ (and so on for any additional samples). If we use the symbol n_T for the total number of scores and <u>53</u> for the number of groups (samples), the formula for the within-groups degrees of freedom is:

$$df_W = 54$$

Putting the sum of squares and the degrees of freedom together, finally, we can write the expression for the within-groups variance estimate, <u>55: symbol?</u> . We must calculate the estimate as the sum of the squared deviations divided by the applicable degrees of freedom, or in symbols:

$$s_W^2 = 56$$

47.

48.

49.

50.

51.

52.

53.

54.

55.

56.

The Among-Groups Quantities

Recall that we are using the samples on hand to construct two estimates of the inherent variation that we assume to be common to the popula-

tions of interest, the variation that is expressed in their variance
57: symbol? . The other estimate of this quantity is called the among-   57.
groups variance estimate, and it is symbolized 58 . It too derives from a   58.
sum of squared deviations, here called the among-groups sum of 59 ,   59.
symbolized 60 . The degrees of freedom for this sum of squares is called   60.
the among-groups degrees of freedom, symbolized 61 .   61.

The among-groups sum of squares is calculated in an unusual way. *For
each score*, find the difference between the mean of its own sample and
the mean of all the scores. The symbol for the mean of a sample is, of
course, 62 , and the symbol for the mean of all the scores, which is   62.
called the 63 mean, is $\overline{\overline{X}}$. Thus for a given score, we must calculate   63.
($\overline{X} - \overline{\overline{X}}$). And of course we must go on to square each such deviation
and add up all the resulting squares. The middle portion of the table on
p. 312 makes the pattern clear. The formula is:

$$SS_A = 64$$ 64.

The degrees of freedom for this sum of squares are one fewer than the
number of 65 . In symbols,   65.

$$df_A = 66$$ 66.

Dividing the sum of squares by the degrees of freedom yields the
among-groups variance estimate. In symbols:

$$s_A^2 = 67$$ 67.

The *F* Ratio

We now have our two estimates of σ^2, the variance we have assumed to
be common to the populations of interest and the result of inherent
variation. To compare the estimates, we divide one by the other, calculating their ratio. In this ratio, the 68 -groups variance estimate always   68.
goes on top (serving as the numerator). The resulting quantity is called
a(n) 69 ratio. The formula, then, is: 69.

$$F = 70$$ 70.

When the treatments we are comparing all have the same effect, so that
the populations all have the same mean and the null hypothesis is
 71: true or false? , the two variance estimates tend to be 72: equal or   71.
unequal? in size. Thus the *F* ratio tends to be about 1.00. (Dividing a   72.
number by the same number produces an answer of 1.)

But of course, even if the null hypothesis were true, we would not expect to get an F ratio of exactly 1.00 in every study, because of the variation that occurs among samples drawn from a given population. Which F ratios are likely and which are unlikely when the null hypothesis is true? The answers come from the distribution of F, which like the distribution of __73__ is actually a family of curves. And like the distribution of __74__, that of F depends on the degrees of freedom for the variance estimates. But with F, because there are two variance estimates involved, there are two values for the df to consider, one associated with the numerator of the ratio, __75: symbol for the df?__, and the other associated with the denominator of the ratio, __76: symbol for the df?__.

    There is a table of critical values of F in Appendix I, and in using this table, then, we must look in a place that depends not only on the level of significance that we choose (.05 or .01), but also on df_A and df_W.

    If our obtained value for the F ratio is larger than the critical value, we have evidence that the null hypothesis is __77: true or false?__, and we can conclude that at least one population mean is probably different from the others. Why? Because getting an unusually large F ratio is, of course, unusual if the null hypothesis is true, but it is the most likely occurrence when the null hypothesis is false.

    And why is a large F ratio likely when the null is false? If the null is false, there are two causes for the variation among the sample means: variation among the population means *plus* the __78__ variation in the populations that is always a factor. Thus the estimate of the inherent variation derived from the variation among the sample means—that is, the among-groups variance estimate—will probably not be a good estimate; it will probably be too large. And in calculating the F ratio, we will probably be dividing something relatively large by something smaller, getting a ratio considerably over 1.00.

73.
74.

75.
76.

77.

78.

Assumptions Underlying the *F*-Test

Underlying the F-test for the one-way analysis of variance are the same assumptions that underlie the __79__-test for __80: dependent or independent?__ samples: homogeneity of variance, normality of the distributions of the populations of interest, independence of the samples, and random sampling.

    As is the case with the t-test, the truth or falsity of the assumptions of normality and homogeneity of variance becomes cause for concern only when samples are __81: large or small?__. The larger the samples, the greater

79.
80.

81.

the departure from homogeneity of variance and normality of shape that can occur without undue distortion of the outcome of the *F*-test.

Other Aspects of the Analysis of Variance

1. *The total sum of squares and degrees of freedom:* The analysis of variance *is* an analysis in the sense that it does decompose something. It decomposes a quantity called the total sum of squares, symbolized __82__ .

This is another sum of squared __83__ , but this one is calculated by subtracting from each score across all samples the __84__ mean $\overline{\overline{X}}$. We must again square each deviation and then sum the deviations, of course. In symbols,

$$SS_T = 85$$

82.
83.
84.

85.

When the arithmetic is correct, the total sum of squares equals the among-groups sum of squares plus the within-groups sum of squares. In symbols, $SS_A + SS_W = SS_T$.

Associated with the total sum of squares is the total degrees of freedom, symbolized __86__ . This quantity is one fewer than the total number of cases across all samples. In symbols,

$$df = 87$$

86.

87.

And just as the among- and within-groups sums of squares add to the total sum of squares, so the among- and within-groups __88__ add to the total degrees of freedom. In symbols, $df_A + df_W = df_T$.

88.

2. *Raw-score formulas:* The formulas for the sums of squares listed above require us to work with deviations between one number and another. In most studies, calculating deviations is laborious, because they turn out to be messy numbers with fractional parts. To save computational labor, we can work entirely with the raw scores. The formulas we need appear in Section 18.7 of your text.

Following-Up a Significant *F* Value

When we obtain a statistically significant *F* value (that is, when we __89: accept or reject?__ the null hypothesis), we draw a statistical conclusion that is vague: it is not true that the population means are all equal. But exactly what is the pattern of differences among them? To answer this question, statisticians have devised a variety of tests. Some of them compare the sample means two at a time, just as a *t*-test would.

89.

These tests are of two types. When the logic of the study, rather than the obtained results, suggests which comparisons we should make, the comparisons are said to be   90: planned or post hoc? . Such comparisons   90.
are legitimate even if the F test yields a nonsignificant result. The text does not present any specific comparisons of this kind.

On the other hand, when we become interested in certain comparisons only after finding a significant F value—that is, when our interest is stimulated by the results we obtained, not by any prior rationale—the appropriate comparisons are called   91 .   91.

A technique for making post hoc comparisons is Tukey's HSD test. HSD stands for   92 . The test applies to samples of   93: equal   92.
or different?   sizes, and it requires us to calculate a value called the critical   93.
HSD. We then compare the difference between a pair of sample means with the critical HSD. If the difference equals or exceeds the critical HSD, we   94: accept or reject?   the null hypothesis that the corresponding   94.
populations have the same mean.

ANNALS OF EGREGIOUS EXAMPLES, *continued*

Look back at the description of the marketing survey offered as an egregious example in Ch. 14 of this *Guide*. In his first study, the businessman really tested five versions of that frozen foodstuff he had been hired to evaluate. He employed not just the two groups of subjects described before, but a total of five groups, one for each kind of food, and he analyzed the data using a series of t-tests for independent samples, comparing the means of the samples of scores two means at a time.

1. You now know that multiple t-tests are inappropriate for a study like this. (See pp. 305-306 of the text.) What inferential technique should the businessman have used on the full set of data from his first study?

In the second study, the perpetrator again tested five versions of a product, but this time he had only one group of subjects, each of whom tasted and rated all five versions. He again analyzed the data using t-tests for independent samples. He thus committed two errors. As you explained if you did the exercise under the heading "Annals of Egregious Examples" earlier in this book, the t-tests should have been those for dependent samples. But that's only the first error. Really, the man should have used not t-tests but an analysis of variance.

The Analysis of Variance for a One-Way Experimental Design

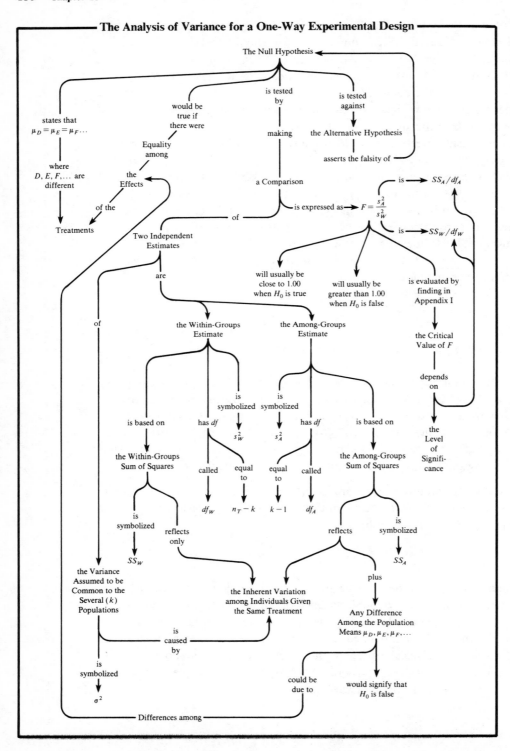

The Null Hypothesis

states that
$\mu_D = \mu_E = \mu_F \ldots$

would be true if there were

is tested by

is tested against

where $D, E, F, \ldots$ are different

Equality among

the Effects

making

the Alternative Hypothesis

asserts the falsity of

of the

Treatments

a Comparison

is expressed as → $F = \dfrac{s_A^2}{s_W^2}$

is → SS_A/df_A

is → SS_W/df_W

of

Two Independent Estimates

are

will usually be close to 1.00 when H_0 is true

will usually be greater than 1.00 when H_0 is false

is evaluated by finding in Appendix I

the Critical Value of F

depends on

the Level of Significance

of

the Within-Groups Estimate

the Among-Groups Estimate

is symbolized

is symbolized

is based on

has df

is symbolized

s_W^2

s_A^2

has df

is based on

the Within-Groups Sum of Squares

called

equal to

equal to

called

the Among-Groups Sum of Squares

df_W

$n_T - k$

$k - 1$

df_A

is symbolized

reflects only

reflects

is symbolized

SS_W

SS_A

the Variance Assumed to be Common to the Several (k) Populations

the Inherent Variation among Individuals Given the Same Treatment

plus

Any Difference Among the Population Means $\mu_D, \mu_E, \mu_F, \ldots$

is caused by

is symbolized

could be due to

would signify that H_0 is false

σ^2

Differences among

2. Is the type of analysis of variance applicable to the first study also suitable for the second study? Why or why not?

There *is* a type of analysis of variance applicable to a design like that of the second study here. It is presented in advanced textbooks.

SELF-TEST

The null hypothesis in the one-way analysis of variance states that ... A) all samples have the same mean. B) all samples have the same variance. C) all populations have the same mean. D) all populations have the same variance.

1.   A B C D

The alternative hypothesis in the one-way analysis of variance ... A) may be either directional or nondirectional. B) specifies exactly how the populations differ. C) specifies exactly how the samples differ. D) states only that the population means are different in some way.

2.   A B C D

Inherent variation ... A) causes the scores within a given sample to vary. B) causes the means of the several samples to vary. C) is reflected in the quantity symbolized σ^2.   D) is all of these things.

3.   A B C D

If the null hypothesis in the one-way analysis of variance is *false*, ... A) s_A^2 is likely to be larger than s_W^2. B) s_A^2 is likely to be about the same size as s_W^2. C) s_A^2 is likely to be smaller than s_W^2. D) homogeneity of variance cannot occur.

4.   A B C D

The degrees of freedom for the among-groups variance estimate are ... A) $n_T - 1$. B) $n_T - k$. C) k. D) $k - 1$.

5.   A B C D

The grand mean is symbolized ... A) X. B) $\bar{X}$. C) $\bar{\bar{X}}$. D) μ.

6.   A B C D

In the analysis of variance, which quantity is analyzed (partitioned)? A) SS_A. B) SS_T. C) SS_W. D) σ^2.

7.   A B C D

An F ratio is based on which of the following degrees of freedom? A) df_A. B) df_T. C) Both of these. D) Neither of these.

8.   A B C D

The assumptions underlying the F-test for the one-way analysis of variance ... A) include $\sigma_D^2 = \sigma_E^2 = \sigma_F^2 = ...$ B) are the same as those for the t-test comparing means of dependent samples. C) are unimportant in practice when sample sizes are small. D) include normality of the distributions of the samples.

9.   A B C D

A social psychologist is studying the effect of audience reaction on the persuasiveness of a speaker. Her subjects watch and hear one of three video tapes of a politician attempting to pursuade an audience that welfare payments should be increased. In the control condition, the subjects hear no reaction from the audience. In a second condition, the subjects hear the audience repeatedly applaud as the politician speaks. In the third condition, the subjects hear the audience repeatedly mutter negatively as the politician speaks. The researcher's theory makes a prediction about how the control condition will differ from the average of the other two conditions, and about how the second and the third condition will differ. Which kind of test should the researcher do? A) Planned comparison.   B) Tukey's HSD test.   C) Post hoc comparison.   D) t-test.

10.   A   B   C   D

CHAPTER 19

ANALYSIS OF VARIANCE: THE TWO-FACTOR DESIGN

OBJECTIVES

Ch. 18 presented a type of analysis of variance that is useful in an investigation that varies the treatment of the subjects in just one way. Ch. 19 now introduces another type of analysis of variance, one that is valuable for an investigation that varies the treatment in two ways simultaneously. In studying Ch. 18, work to understand the following symbols, terms, concepts, principles, and techniques:

The Factorial Design
    Factor Consisting of Two or More Levels
        Row Factor   •   Column Factor
    Cell: a Combination of a Row Level and a Column Level
    Advantages over the One-Factor Design
        Conclusions Can Be Broader
        Efficiency is Greater
        Interaction between the Two Factors Can be Studied
            Interaction as a Cause for Caution in Interpreting the Effects of a Factor
The Factorial Analysis of Variance
    Cell Means   •   Row Means   •   Column Means
    The Main Effects:
        The Main Effect of the Row Factor
            The Null Hypothesis   •   The Alternative Hypothesis
        The Main Effect of the Column Factor
            The Null Hypothesis   •   The Alternative Hypothesis
    The Interaction between the Row and Column Factors
        The Null Hypothesis   •   The Alternative Hypothesis
The Partition of the Total Sum of Squares
    The Total Sum of Squares: $SS_T = \Sigma(X - \bar{\bar{X}})^2$
    The Raw-Score Formula
    The Within-Cells Sum of Squares: SS_{WC} Based on Squared Deviations of the Kind $(X - \bar{X}_{cell})^2$

Also in studying Ch. 19, work to learn these skills:

Execution of the Analysis of Variance for a Two-Factor Design, Given Raw Scores in Equal-Sized Samples
>   Statements of the Null and Alternative Hypotheses
>   Computation of:
> SS_T and df_T • SS_{WC} and df_{WC} • SS_R and df_R • SS_C and df_C
>   • $SS_{R \times C}$ and $df_{R \times C}$ • s^2_{WC} • s^2_R • s^2_C • $s^2_{R \times C}$
>   • F for the Row, Column, and Row by Column Effects
>   Use of Appendix I to Determine the Critical Values of F
>   Decisions on the Null Hypotheses
>   Interpretation of the Results
>   Computation of Cell and Marginal Means

REVIEW

Look at Table 19.1 on p. 331. Ch. 19 is concerned with studies of the kind illustrated in this table. The table diagrams a hypothetical experiment in which a researcher has observed subjects performing a verbal learning task. (It might be something like learning to associate one nonsense syllable with another—learning that *dak* and *yug* go together, for example, and similarly for *zid* and *lak* and many other such pairs.) The researcher studied a total of __1__ subjects.

  1.

  A third of the subjects performed their task under instructions of a kind called Type A, which put no pressure on the subject. The number of such subjects, n_A, was __2__ . A second third of the subjects performed the task under instructions of Type B, which applied a moderate degree of pressure to do well; $n_B = $ __3__ . The remaining third of the subjects performed the task under instructions of Type C, which applied strong pressure to do well; $n_C = $ __4__ . These three kinds of circumstance are called treatments or conditions, and they make up a treatment variable. The treatment variable here is Instructions, and each type of instructions is represented by a different __5: row or column?__ in the table. In statistical terminology, a treatment variable like Instructions is called a __6__ , and the specific treatments or conditions are called the __7__ of the __8__ . Because the table illustrates the three levels of this factor as columns, the factor is called the __9__ factor in the study.

  2.

  3.

  4.

  5.

  6.

  7.

  8.

  9.

  If this were all there were to the study—if there were just this one treatment variable—the study would have the kind of design considered

in the previous chapter, a __10__ design. But the subjects who received each   10.
type of instruction (the subjects represented in each __11: row or column?__   11.
of the table) were of two types. One type was highly prone to anxiety,
and the other type was not. Of the 20 subjects who received each type of
instructions, there were __12: how many?__ of the high-anxiety type and   12.
__13: how many?__ of the low-anxiety type. This study thus has a second   13.
factor, which is Type of Subject. (We might also call it Anxiety.) The
factor has __14: how many?__ levels, namely high and low, and because the   14.
table illustrates each level of the factor as a row, Type of Subject serves
as the __15__ factor in the study.   15.

A study of this sort is said to have a __16__ design. The essence of the   16.
design is this: each level of one factor appears in combination with each
level of the other factor. In the example, each level of the factor
Instructions appears in combination with each level of the factor Type of
Subject. That is, each type of instructions (each column) appears once in
combination with one type of subject (one row) and again in combination with the other type of subject (the other row). The design of the
study that serves as the example is called a 2×3 ("two by three")
factorial design. The "2" is the number of levels of the __17__ factor, and   17.
the "3" is the number of __18__ of the column factor.   18.

Each combination of a row level and a column level is called a __19__. In   19.
the example, there are __20: how many?__ cells, and this number is the   20.
product of the little multiplication problem that describes the design:
$2 \times 3 = 6$, the number of cells.

In the example, all cells contain the same number of cases, which is
__21__. Unequal cell sizes in a factorial design can cause problems in the   21.
analysis of the data, and this chapter considers only situations in which
there are equal numbers of cases per __22__.   22.

This chapter also considers only situations in which the cells contain
__23: what kind of?__ groups.   23.

Means

At the conclusion of a study of this kind, one should describe the data by
calculating the various means shown in Table 19.2 at the bottom of p.
331. The six means within the body of the table are called __24__ means;   24.
each is symbolized $\overline{X}$. The mean of all scores in a given row (in the
example, there are __25: how many?__ scores per row) is called a __26__ mean.   25.
The mean of the scores in the top row, the mean for all subjects prone to   26.
high anxiety, is 84 and is symbolized __27__. Note that a number like this is   27.

not merely the mean of the scores in its row but also the mean of the cell means in its row. This will be true for all the row means, but only when there are   28: equal or unequal?   numbers of scores per cell.   28.

The mean of all the scores in a given column (in the example, there are   29: how many?   scores per column) is called a   30   mean. The mean for   29.
the 20 subjects in the left-hand row, the mean for the subjects who   30.
worked under instructions of Type A, is 85; it is symbolized   31 . As for   31.
the rows, when the cells contain samples of equal sizes, the mean of all
scores in a column is also the mean of the   32   in that column.   32.

In the lower right corner, outside the body of the table, there appears
the   33   mean, which is the mean of all the scores. And because the cell   33.
sizes are all equal here, the grand mean is also the mean of the row means
and the mean of the column means and the mean of the cell means.

Each collection of scores in a given cell of this study is, of course, only
a sample from a population. Such a population consists of the scores
characterizing all conceivable subjects like those actually studied who
could be observed under the circumstances that applied to the given cell
of the design. (In the upper-right cell of the design, for instance, the 10
scores were a sample from the population of the scores that would be
earned if we ran through this experiment all possible subjects of whatever
nature we studied who are prone to high anxiety, testing them for their
verbal-learning performance under instructions of Type C.)

Corresponding to each sample mean in the table at the bottom of p.
331, there is a population mean. (The population means are shown in the
map for this chapter. The terminology there is like that of Table 19.3 on
p. 337: the rows are called R_1 and R_2 instead of H and L, and the
columns are called C_1, C_2, and C_3, instead of A, B, and C.)

Main Effects

As noted above, if the study illustrated in Tables 19.1 and 19.2 had
examined the effect of only one factor—Type of Instructions, say—it
would have the one-way design that was the topic of the previous
chapter. In such a case we would analyze the data with a one-way   34 ,   34.
and the null hypothesis would be an assertion about the three populations of scores that we are envisioning in each column of our design.
(One population consists of scores generated under instructions of Type
A, a second consists of scores generated under instructions of Type B,
and a third of scores generated under Type C.) The null hypothesis
would assert that   35: symbols? . Evidence on the plausibility of this null   35.

would come, in part, from the variation among the corresponding sample means, which in the table at the bottom of p. 331 are symbolized _36_ .   36.

In a factorial study, it is still possible to analyze the data in this way. The question about the effect of the column factor becomes, however, a question about its *overall* (*average*) influence: what was the effect of the Type of Instructions *over all* levels of the other factor—that is, what was the effect of the Type of Instructions *averaged* across the levels of the other factor? (Here the averaging is over the two levels of Type of Subject.) The overall (average) influence of a factor is called a _37_ . Thus   37. the main effect of Instructions is the difference (if any) among the means of the verbal-learning scores under the three types of instructions when we average over both kinds of subject.

The null hypothesis for that aspect of the factorial analysis of variance that tests a main effect is always an assertion about the population means corresponding to the sample means that appear along one margin of a table like 19.2. For the column factor in the present example, the null hypothesis, in symbols, asserts that _38_ . The alternative hypothesis states   38. that the null is false. As in a one-way analysis of variance, this alternative hypothesis is deliberately vague, covering all of the many possible ways in which the null could be false.

A factorial design with two factors permits us to ask not only about the main effect of the column variable but also about the main effect of the _39_ . This main effect, like the other, is an overall effect. It is the   39. effect of the row factor averaged over all levels of the _40_ . In the   40. example, the main effect of Type of Subject is the difference in mean verbal-learning performance between high-anxiety and low-anxiety subjects when we average over all three _41_ of the column factor (over all   41. three types of instructions). The null hypothesis for the main effect of the row factor in the example asserts, in symbols, that _42_ . Again, the   42. alternative hypothesis simply says that the null is _43_ .   43.

Interaction

A factorial design is not merely a way of assessing two main effects at once. It permits us to ask an entirely new question: is there a(n) _44_   44. between the two factors?

There is *no* interaction between two factors when the effect of one factor is the same at each level of the other factor. For example, there would be no interaction between Type of Subject and Type of Instructions if the effect of Type of Subject—the difference between high-anxiety and low-anxiety subjects—were the same under each type of instructions. To be even more specific in this example, there would be no interaction between these two factors if the mean score for high-anxiety subjects were smaller than the mean for low-anxiety subjects by a certain amount under instructions of Type A *and* smaller than the mean for low-anxiety subjects by the *same* amount under instructions of Type B *and again* smaller by the *same* amount under instructions of Type C.

This state of affairs is illustrated in Figure 19.2 on p. 335. The three differences under consideration are represented by the double-headed arrows labeled (1), (2), and (3). Note. that a question of interaction is a question about *differences* between means: is the difference between the two means for the two levels of the row factor the same at each level of the 45 factor? In asking about interaction, we are not asking about the value of any one mean. It is irrelevant to the issue of interaction that the low-anxiety subjects under instructions of Type B performed better, for example, than the low-anxiety subjects under instructions of Type C in the hypothetical outcome illustrated in the figure on p. 335. A question of interaction is a question about the *effect* of one variable at the various levels of the other variable, and a question about the effect of a variable is a question about the pattern of differences among the means of the scores at the various levels of this variable.

45.

In determining whether an interaction is present, we can ask either of two questions and get the same answer. The question we have been asking here is whether the effect of the row factor is the same at each level of the column factor. But we could also ask, Is the effect of the column factor 46 at each level of the row factor? In the example, we could ask whether the effect of Type of Instructions is the same at each level of the factor called Type of Subject. That is, is the pattern of differences among the means for the three types of instructions the same for high-anxiety subjects as it is for low-anxiety subjects? If you inspect the figure on p. 335 carefully, you will see that the answer to this other question is the same as the answer to the first question, namely 47 . Thus both ways of asking the question indicate no interaction for the hypothetical pattern of results shown in this figure.

46.

47.

An interaction between two factors *is* present when the effect of either is *not* the same at each level of the other. Look at Figure 19.1 on p. 333 for an illustration of this state of affairs.

In the discussion of interaction thus far in this *Guide*, we have not distinguished between samples and populations. Sample means may or may not show an interaction; population means may or may not show an interaction. Our concern is with the population means, of course, and the factorial analysis of variance permits us to test the null hypothesis that in the populations there is _48: an or no?_ interaction. The alternative hypothesis states that in the population there is _49: an or no?_ interaction.

48.

49.

It is important to determine whether there is an interaction between two factors, because when interaction is present, we cannot take a main effect at face value. Considering a main effect by itself may give a misleading impression when the factor that shows the main effect interacts with the other factor. The results illustrated in Table 19.2 on p. 331 and in Figure 19.1 on p. 333, for example, include a main effect for Type of Subject: high-anxiety subjects did _50: better or worse?_ than low-anxiety subjects ($\overline{X}_H$ = _51_ while $\overline{X}_L$ = _52_). But this assertion is true only overall, true only on the average. Type of Subject interacts with Type of Instructions, and the high-anxiety subjects did worse than their low-anxiety counterparts only under instructions of Types B and C, with moderate or strong pressure to do well. The suggestion in the main effect that high-anxiety subjects are less adept at verbal learning is wrong under instructions of Type A. Here there is _53: an or no?_ important difference between the two types of subject.

50.

51.

52.

53.

(Studying the learning of associations between nonsense syllables may have seemed a waste of time when you first heard about it, but perhaps now you can appreciate that it provides a good way to investigate important matters, such as what kind of atmosphere permits the best learning for various kinds of student. Because the nonsense syllables are equally novel to all subjects, no subject has an unfair "head start" in learning what there is to learn in the experiment.)

The Logic of the Factorial Analysis of Variance

We have now reviewed the three null hypotheses to which the factorial analysis of variance is addressed. One asserts that in the _54: populations or samples?_ there is no main effect for the row factor; a second asserts that in the populations there is no main effect for the _55_ ; and the third asserts that in the populations there is no _56_ between the two factors. To test these null hypotheses, the analysis generates three _57_ -tests, one for each null. As in the one-way analysis of variance, each F ratio compares one estimate of a population variance with another such estimate. When the null hypothesis is true, the two estimates tend to be about the same size, and the F ratio is about 1.00. When the null hypothesis is false, the variance estimate that serves as the numerator is

54.

55.

56.

57.

likely to be   58: larger or smaller?   than the other, and the F ratio is thus   58.
likely to be larger than 1.00.

The variance estimated by the quantities that make up the F ratio is
the 59 variation, and as in the 60 analysis of variance, we assume that   59.
it is the same for all populations of interest—which are the populations   60.
of scores for each cell of the study.

In a one-way analysis of variance, the quantity that provides an estimate
of inherent variation free from any differential effect of the factor under
study is the 61 -groups variance estimate, symbolized 62 . In a two-way   61.
(factorial) analysis, there is an analogous quantity, and it is the 63   62.
variance estimate, symbolized 64 .   63.
                                                                                 64.

In a one-way analysis, the variance estimate other than the within-
groups estimate measures the inherent variation *plus* any differential
effect of the one factor under study. This other variance estimate is the
 65 -groups variance estimate, symbolized 66 . Corresponding to this   65.
quantity in a two-way factorial analysis, there are 67: how many?   vari-   66.
ance estimates. One is the variance estimate for the row factor, and it   67.
measures inherent variation plus any 68 effect of the row factor in the   68.
populations; its symbol is 69 . A second such quantity is the variance   69.
estimate for the column factor, and it measures inherent variation plus
any main effect of the column factor in the populations; its symbol is
 70 . The third quantity analogous to the among-groups variance esti-   70.
mate, s_A^2, is the variance estimate for the interaction between the row and
the column factor, and it measures inherent variation plus any interaction in the populations; its symbol is 71 .   71.

The Partition of the Total Sum of Squares

In a factorial analysis of variance, as in a one-way analysis of variance, a
variance estimate, symbolized s^2, is calculated as a sum of squared
deviations, called a sum of 72 , divided by the associated 73 . And as in   72.
the one-way analysis, the total sum of squares is partitioned, but here   73.
there are 74: how many? components.   74.

The total sum of squares is based on the squared deviations of an
individual score from the 75 mean of all scores in the samples on hand.   75.
Quantities of this kind are described by the formula 76 , and the total   76.
sum of squares is symbolized 77 .   77.

The within-cell sum of squares is based on the squared deviation of an
individual score from the mean of its own 78: row, column or cell? .   78.
Quantities of this kind are described by the formula 79 , and the   79.
within-cells sum of squares is symbolized 80 .   80.

The row sum of squares is based on the squared deviation of a row mean from the grand mean. For a design with two rows, symbolized R_1 and R_2, the squared deviations would be described by the formulas $(\overline{X}_{R_1} - \overline{\overline{X}})^2$ and $(\overline{X}_{R_2} - \overline{\overline{X}})^2$. (Remember that one such squared deviation must be calculated for each raw score in the analysis, so the summation that adds up the various quantities of the two kinds just named is a summation over all scores across the several samples on hand.)

The column sum of squares is based on the squared deviation of a column mean from the __81__ mean. For a design with three columns symbolized C_1, C_2, and C_3, the squared deviations would be described by the formulas $(\overline{X}_{C_1} - \overline{\overline{X}})^2$, $(\overline{X}_{C_2} - \overline{\overline{X}})^2$, and $(\overline{X}_{C_3} - \overline{\overline{X}})^2$. (Remember again that one such squared deviation must be calculated for each raw score in the analysis.)   81.

The interaction sum of squares is based on the squared deviations of the __82: row, column, or cell?__ means from those means that would be predicted to occur if there were no interaction in the population. Its symbol is __83__.   82.

83.

These four sums of squares add to the total: $SS_T = SS_{WC} + SS_R + SS_C + SS_{R \times C}$.

The Partition of the Total Degrees of Freedom

Associated with each sum of squares is a degrees of freedom. The total degrees of freedom is __84: how many?__ less than the total number of scores across all the samples. In studies of the kind under consideration, with __85: equal or unequal?__ numbers of scores per cell, the total number of scores can be calculated as follows: Multiply the number of rows, symbolized __86__, by the number of columns, symbolized C. That yields the number of __87: rows, columns, or cells?__. Multiply the result by the number of scores per cell, whose symbol is n_{cell}. The whole thing with the "minus one" at the end looks like this: $df_T = $ __88__.   84.

85.

86.

87.

88.

The degrees of freedom for each main effect are calculated in a fashion analogous to the calculation of the degrees of freedom for the among-groups quantities in the one-way analysis. In the one-way analysis, the one factor has a certain number of levels (also called treatments or groups), which is symbolized k, and the degrees of freedom for the associated sum of squares is __89__. In a two-way analysis the row factor has a certain number of levels symbolized R, and the degrees of freedom for the row sum of squares are $df_R = $ __90__. Similarly, given a column   89.

90.

factor with C levels, the degrees of freedom for the column sum of squares are $df_C = \underline{\ 91\ }$.   91.

For the interaction sum of squares, the degrees of freedom are the $\underline{92:\ \text{sum or product?}\ }$ of those for the two main effects. That is,   92.
$\overline{df_{R \times C}} = \underline{\ 93\ }$.   93.

Again the parts add to the whole: $df_T = df_{WC} + df_R + df_C + df_{R \times C}$.

Raw-Score Formulas

In Ch. 19, the text shows you how to calculate the sums of squares only in terms of raw scores. No formulas involving squared deviations appear, because these are highly complex and would be of little use in practical work.

Concluding the Analysis of Variance

Having calculated the necessary sums of squares and the associated degrees of freedom, we divide a sum of squares by its degrees of freedom to obtain a variance $\underline{\ 94\ }$. The formulas appear on p. 342 of the text. The   94.
calculations are conveniently summarized in a table like that on p. 343. As for the one-way analysis, we must obtain $\underline{\ 95\ }$ values of the F ratio   95.
from Appendix I, and a critical value depends on three things: the level of significance that we choose (.05 or .01), the $\underline{\ 96\ }$ for the numerator of   96.
F ratio, and the degrees of freedom for the denominator of the F ratio. The latter will always be the degrees of freedom for the $\underline{\ 97\ }$ variance   97.
estimate. If an F ratio equals or exceeds the critical value for whatever level of significance we choose, we $\underline{\ 98\ }$ the null hypothesis tested by our   98.
F ratio.

Underlying these F tests are the same assumptions that we encountered in the previous chapter: independence of the samples in the several cells, random sampling, $\underline{\ 99\ }$ of variance, and $\underline{\ 100\ }$ of the populations in   99.
the cells. As usual, the latter two assumptions can be wrong without   100.
seriously affecting the validity of the test if sample sizes are
$\underline{101:\ \text{large or small?}\ }$.   101.

To understand the data, it is essential to do more than just calculate the F ratios and accept or reject the associated $\underline{\ 102\ }$ hypotheses. We   102.
should compute the means for the cells, the rows, and the columns, and we should display them in a table or figure like those in this chapter. We can compare pairs of means using a variation of the $\underline{\ 103\ }$ test that was   103.
introduced in the previous chapter.

The Factorial Analysis of Variance for a Two-Factor Design

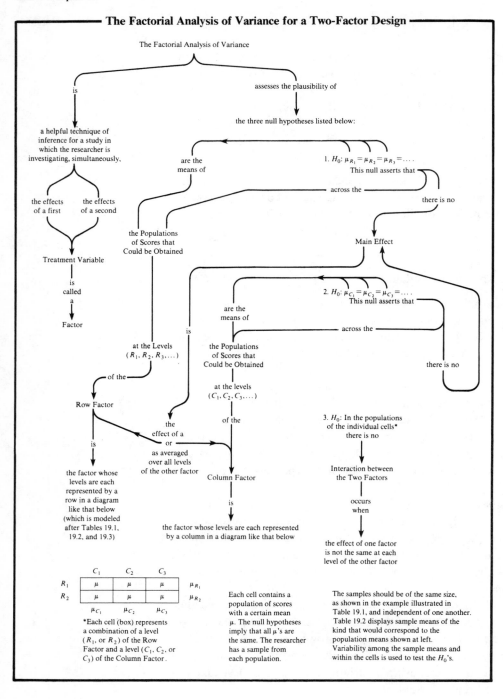

The Factorial Analysis of Variance

is → a helpful technique of inference for a study in which the researcher is investigating, simultaneously,

the effects of a first   the effects of a second

Treatment Variable

is called a

Factor

assesses the plausibility of → the three null hypotheses listed below:

1. H_0: $\mu_{R_1} = \mu_{R_2} = \mu_{R_3} = \ldots$.
This null asserts that → there is no → Main Effect → across the → there is no

are the means of → the Populations of Scores that Could be Obtained

at the Levels $(R_1, R_2, R_3, \ldots)$

of the → Row Factor

is → the factor whose levels are each represented by a row in a diagram like that below (which is modeled after Tables 19.1, 19.2, and 19.3)

the effect of a or as averaged over all levels of the other factor

2. H_0: $\mu_{C_1} = \mu_{C_2} = \mu_{C_3} = \ldots$.
This null asserts that → there is no → across the

are the means of → the Populations of Scores that Could be Obtained

at the levels $(C_1, C_2, C_3, \ldots)$

of the → Column Factor

is → the factor whose levels are each represented by a column in a diagram like that below

3. H_0: In the populations of the individual cells* there is no

Interaction between the Two Factors

occurs when

the effect of one factor is not the same at each level of the other factor

	C_1	C_2	C_3	
R_1	μ	μ	μ	μ_{R_1}
R_2	μ	μ	μ	μ_{R_2}
	μ_{C_1}	μ_{C_2}	μ_{C_3}	

*Each cell (box) represents a combination of a level $(R_1, \text{ or } R_2)$ of the Row Factor and a level $(C_1, C_2, \text{ or } C_3)$ of the Column Factor.

Each cell contains a population of scores with a certain mean μ. The null hypotheses imply that all μ's are the same. The researcher has a sample from each population.

The samples should be of the same size, as shown in the example illustrated in Table 19.1, and independent of one another. Table 19.2 displays sample means of the kind that would correspond to the population means shown at left. Variability among the sample means and within the cells is used to test the H_0's.

Advantages of a Factorial Design

A study with a two-way factorial design enjoys three advantages over any study with the simpler one-way design. First, any conclusion about either of the two factors can be broader than the corresponding conclusion about the one factor studied in a one-way design. In a *one*-factor study investigating only the effect of type of subject (prone to high or prone to low anxiety) on learning ability, for example, all subjects would be tested under a given set of instructions, and any conclusion would be limited to performance under just that one set. But in the two-factor study that serves as an example in this chapter, we can draw a broader conclusion, one that applies to 104: how many? different types of instructions.   104.

The second advantage lies in the increased efficiency of the design. Because we are studying 105: how many? factors at once, we are answering two questions for the price of one: one question about the effect of one factor, and another question about the effect of the other factor.   105.

The third advantage lies in our ability to inquire about the 106   106. between those two factors.

SELF-TEST

Suppose an educational psychologist is carrying out an experiment in which grade-school children hear one of four stories: a stereotypical (and sexist) story in which a boy is the active hero while a girl just passively watches him, a reverse-stereotype story in which a girl is active while a boy just watches, a story in which both a boy and a girl are active, or a story in which both a boy and a girl are passive. Of the children who hear a given story, 15 are male and 15 female. After hearing the story (whichever of the four kinds it is), each subject is given an opportunity to work at the mildly frustrating task of using long-handled spoons to rearrange a terrarium in a narrow-necked jar. The psychologist observes how long the subject persists at the task. This experiment has which kind of design?   A) A one-way design in which type of story is the treatment variable.   B) A one-way design in which sex of subject is the treatment variable.   C) A two-way design in which type of story and sex of subject are the treatment variables.   D) A two-way design in which type of story and persistence at the task are the treatment variables.   1.   A   B   C   D

In the experiment described in Question 1, how many cells are there? A) 2   B) 4   C) 6   D) 8   2.   A   B   C   D

In the experiment described in Question 1, how many levels does the variable Type of Story have?   A) 2   B) 4   C) 6   D) 8

3.   A   B   C   D

In the experiment described in Question 1, the mean score for all boys can be compared with the mean score for all girls. This comparison provides evidence on which of the following questions?   A) Is there a main effect for type of story?   B) Is there a main effect for sex of subject?   C) Is there any interaction between type of story and sex of subject?   D) Is there homogeneity of variance in the populations?

4.   A   B   C   D

Which of these outcomes for the experiment described in Question 1 would indicate interaction between type of story and sex of subject for the samples on hand?   A) The mean persistence score for all boys equals the mean persistence score for all girls.   B) The mean persistence score for all children who hear a given story is the same for all four stories.   C) The mean persistence score for all boys is 18.3 seconds longer than the mean persistence score for all girls.   D) The mean persistence score for boys is greater than that for girls when the story the child just heard featured an active male character, but it is lower than the mean persistence score for girls when the story featured an active female character.

5.   A   B   C   D

In a two-way factorial analysis of variance, which of these quantities is a part of the total sum of squared deviations?   A) SS_R B) SS_A C) SS_W D) None of these.

6.   A   B   C   D

In the experiment described in Question 1, what are the degrees of freedom for the factor Sex of Subject?   A) 1   B) 2   C) 3   D) 4

7.   A   B   C   D

In the experiment described in Question 1, what are the degrees of freedom for the interaction between the two factors?   A) 1   B) 3   C) 7   D) 8

8.   A   B   C   D

In a two-factor analysis of variance, which quantity is most similar to the among-groups sum of squared deviations in a one-way analysis?   A) SS_{WC} B) SS_T C) $SS_{R \times C}$ D) SS_C

9.   A   B   C   D

The total degrees of freedom for the experiment described in Question 1 are ...   A) 14.   B) 59.   C) 119.   D) 120.

10.   A   B   C   D

CHAPTER 20

THE BINOMIAL TEST

OBJECTIVES

In previous chapters, you learned techniques of description and techniques of inference suitable for observations that are scores. Ch. 20 now applies the logic of hypothesis testing to studies in which the observations are categorizations and there are only two possible categories. In studying the chapter, work to learn these symbols, terms, concepts, principles, and techniques:

Dichotomous Observations
P: the Probability of One of the Two Possible Results for a Single Dichotomous Observation
   $1-P=$ the Probability of the Other Result
Sampling Experiment Consisting of n Trials
   X: the Count of the Observations of a Given Kind over the n Trials
The Binomial Distribution: the Theoretical Probability Distribution of X
   As a Family of Distributions that Vary Depending on P and n
   Appendix K: a Listing of the Binomial Distribution for $P=.25$ or $.50$ and $n=1-15$
Hypothesis Testing with the Binomial Distribution
   The Null Hypothesis
   The Alternative Hypothesis
      Nondirectional • Directional
   The Level of Significance (α)
   Determination of Sample Size (n)
   Calculation of X, the Count of the Observations of the Kind of Interest
   The Region(s) of Rejection
      The Effective Level of Significance • A Difficulty when the Test is Two-Tailed and $P\neq.50$
   Decision on the Null Hypothesis
Assumptions Underlying the Test:
   Independence of Trials • Constancy of P Across Trials

Also in studying Ch. 20, work to learn these skills:

Use of Appendix K to Determine, Given $P=.25$ or $P=.50$ and n 15 or less:
   The Probability of X Occurrences of a Given Kind over the n Trials

The Probability of More than X Occurrences of a Given Kind over the n Trials
The Probability of Fewer than X Occurrences of the Given Kind over the n Trials

Execution of a Test of the Hypothesis that $P=.25$ or $P=.50$, Given 15 or Fewer Observations:

Statement of the Null Hypothesis • Statement of the Alternative Hypothesis • Choice of the Level of Significance • Calculation of X • Determination of the Region(s) of Rejection • Calculation of the Effective Level of Significance • Decision on the Null Hypothesis

REVIEW

Ch. 20 is devoted to a technique of inference applicable to research in which each observation can be one of only two kinds. In a study aimed at determining whether rats prefer light or darkness, for example, we might give each rat a choice between a path that leads to a bright light and a path that leads to a dim light, and we would classify each rat into one of two categories: "turned toward the light" or "turned toward the darkness." Observations like these that can take only two possible values are called __1__ data.

1.

Two Kinds of Probability that Apply to Dichotomous Data

A researcher who collects dichotomous data is often concerned with the probability of the occurrence of one of the two kinds of observation. Someone wondering whether rats prefer darkness, for example, would want to know if the probability that a rat will turn into the path leading to darkness is greater than .50. The symbol for such a probability is __2__.

2.

Because there are only __3: how many?__ possible kinds of observation, the probability of the occurrence of the other kind is __4__.

3.

4.

We can think of the probability of the occurrence of an observation of a given kind as a proportion: the proportion of observations of that kind in an infinitely large population of dichotomous observations. (In our example, the probability that a rat will turn into a path leading to darkness is the proportion of observations of the type "chose the darker path" in the population of observations that one could generate by extending the study indefinitely.)

What will happen if we collect a sample of observations and count the number that fall into one of the two possible categories? We might find that none of the observations are of the kind of interest; we might find that all of the observations are of this kind; or we might get any count in between. If we employ the usual symbol for the total number of observations in a sample, which is _5_, then, the count of the given kind can vary between _6_ and _7_.

If we know the probability of the occurrence of the observation of interest, it is possible to know the probability that *in our sample* we will obtain a certain number of such observations.

For an example, look at Figure 20.1 on p. 355. This figure tells us some things about the situation in which an observation of a certain type has a probability of _8_ —which is to say that in the population, _9: what fraction?_ of the observations are of this type. (The type is "right" for the hypothetical study that this figure illustrates; the other possible type is "wrong," and the study is one in which a person attempts to identify a card using telepathy.) If we collect four observations (that is, if we "run four trials," drawing a sample of size _10_ from the population), the number that are of a given kind can vary between _11_ and _12_. The possible counts are shown along the *X*-axis in the figure. The left-hand bar, the bar for the count of zero, rises to a height of about one-third (.3164) on the scale of relative frequencies that constitutes the *Y*-axis in the diagram. What does this mean? It tells us that if we were to draw samples of size _13_ over and over from the population, continuing indefinitely, in _14: what proportion?_ of these samples, we would get exactly zero observations of the kind of interest to us.

The figure also indicates that in _15: what proportion?_ of the samples, we would obtain exactly one observation of the kind of interest. The count that would turn up least often in the long run, if we were to collect samples of size four from this population in which one-quarter of the observations are of the kind we're keeping track of, is 4; it would occur with a relative frequency of _16_.

Each of the relative frequencies shown in Figure 20.1 is a probability, for a probability is nothing other than the proportion of times something occurs in the long run. (See p. 156 for a review of the concept of probability.) Thus the probability is _17_ that in any one sample of size four we will obtain zero observations of the kind of interest. And the probability is .4219 that in any one sample of size four we will obtain exactly _18: how many?_ observation of the kind of interest.

Remember that these probabilities are for the case in which the type of observation we're concerned with has a probability of _19_. That is, these probabilities are for the case in which this type of observation constitutes one-fourth of the entire _20_ of dichotomous observations. And these are the probabilities for the case in which sample size is 4.

5.

6.

7.

8.

9.

10.

11.

12.

13.

14.

15.

16.

17.

18.

19.

20.

Don't be confused because we're now talking about two kinds of probability. There is first of all the probability that when we go to make any *one* observation, it will be of a given type. (This is, for example, the probability that when we observe any *one* choice between two paths, the subject will choose the darker one.) This probability can be regarded as the proportion of the observations that are of the type of interest in the 21: sample or population? . The other kind of probability applies to the 21.
outcome of a sample of a particular size. If we draw a sample of some size, what is the probability that the outcome will be a certain number of observations of the type we're counting? In the study illustrated in the figure on p. 355, the sample size is 22 , and thus the number of 22.
observations of the type we're counting can vary from 23 to 24 . As we 23.
have seen, for each possible outcome of the sampling experiment, there is 24.
a certain probability.

Sampling Distributions for Dichotomous Data

The figure on p. 355 is the 25: what kind of graph? for a random 26   25.
distribution. Like any other such distribution, this one consists of the 26.
values of a certain statistic characterizing an infinitely large collection of samples, each of a certain size, each drawn at random from a certain population. The statistic is the 27 of the observations of a given type 27.
within the sample; the sample size is 28 ; and the population is one in 28.
which the proportion of observations of the given type, symbolized P, is
 29 . The figure shows the relative frequency with which each possible 29.
value of the statistic will occur. Here the possible values are 30 . The 30.
relative frequency for a given value of the statistic is also the probability that any one sample will be characterized by this value.

You may find it helpful to compare this sampling distribution with the kind you first met on p. 177. There, the sampling distribution consisted of values of the means of an infinitely large collection of samples of size 25 all drawn from a population of birthweights. The mean of the population was 126 ounces. The figure on p. 176 illustrates the sampling distribution; compare it to the one on p. 355.

A sampling distribution of the kind shown on p. 355 depends on two things: the probability P (the 31 characterizing the population) and the 31.
sample size n. Change either, and the sampling distribution changes. In the same way, the random sampling distribution of the mean depends on the mean of the population, 32: symbol? , and on the sample size. See p. 32.
178 to review.

The family of distributions of the kind shown on p. 355 is called the
__33__ distribution.   33.

Binomial Probabilities

Where did the probabilities shown in the figure on p. 355 come from? If
you remember the principles for calculating probabilities presented in
Ch. 9, you can determine them yourself, as pp. 356-357 show. A shortcut
is to use a formula called the binomial equation, which you may have
studied in a mathematics course. The text does not present the binomial
equation, but it does offer a very useful table listing some of the
probabilities specified by this equation. The table is Appendix K.

Appendix K is for either of two situations. In one, the probability of
getting an observation of a particular type *on any one trial* is .25; in the
other, the probability of getting an observation of a particular type *on
any one trial* is __34__ .   34.

In either situation, the number of trials (the sample size, symbolized
__35__), can be as small as 1 or as large as __36__ .   35.
                                                                        36.
The probabilities that the table lists are for the various possible
outcomes of a study in which we draw a sample of a given size (or "run"
a given number of "trials"). An outcome is a certain number of observations of the type whose probability *on any one trial* is .25 or .50. The
number is symbolized __37__ in the table.   37.

Remember that in previous chapters, this symbol stood for a raw score,
which was something like the birthweight of a baby; it could have taken
any of many possible values, depending on just what was measured. (If X
stands for a birthweight measured to the nearest tenth of an ounce, it can
take values like 118.1, 118.2, 118.3, and so on.) But here, X stands for a
count, not a score. It stands for the count of the observations of the kind
of interest that turned up in a sample of a particular size (symbolized
__38__). As a count, X must be 0 or 1 or 2 or 3, and so on.   38.

We can use Appendix K to determine not merely the probability that
X takes a given value, but also the probability that X takes any of a range
of values. That is, we can use the table to answer a question of the
following form: Suppose the probability of getting an observation of a
certain kind on any one trial is such-and-such. (The probability,
__39: symbol?__ , must be .25 or __40__ if we are to use Appendix K.) Suppose 39.
further that we run so-many trials. (The number of trials, __41: symbol?__ ,   40.
must lie between 1 and __42__ if we are to use Appendix K.) What, then, is   41.
the probability that in our sample of this many trials, there will be *more*   42.

than a certain number of observations that are of the type of interest? Or *less than* a certain number of observations that are of the type of interest?

For an example, look at the problem worked on pp. 358-359. Here $P =$ _43_ and $n =$ _44_. The question is about the probability of getting more than four observations of the kind of interest. This is the probability of obtaining 5 *or* 6 *or* 7. To calculate this probability, we can use the OR/addition rule that Ch. 9 offered on p. 160. The rule is applicable here because the events we are asking about are _45_ exclusive. (See p. 160 to review this concept.) Thus we need only add up the probability of getting exactly 5 observations of the kind in question, the probability of getting exactly 6, and the probability of getting exactly 7. The figure on p. 358 illustrates the calculation.

43.

44.

45.

Hypothesis Testing with the Binomial Distribution

In the great majority of cases in which we collect dichotomous observations, we do not, of course, know the probability of getting an observation of a certain kind on any one trial. In such cases, it is helpful to test a hypothesis about this probability. The logic is _46: the same as or different from?_ that for any other instance of statistical hypothesis testing.

46.

The first step is to state the hypothesis to be tested. As usual, it is an assertion about a population (a probability is a proportion characterizing a population), and it is called the _47_ hypothesis. If we are to use Appendix K, the hypothesis must specify that the probability of interest, symbolized _48_, is .25 or .50.

47.

48.

We must also state an alternative to the null. The alternative hypothesis, symbolized _49_, may be nondirectional or directional. If nondirectional, the alternative says that P _50: symbol?_ .25 or .50 (whichever one the null specifies). There are _51: how many?_ possible alternative hypotheses of the directional kind. One says that $P > .25$ or $P > .50$ (depending on whether null specifies .25 or .50). The other kind of alternative hypothesis says that P _52: symbol?_ whatever value the null states.

49.

50.

51.

52.

Also in our first step, we must choose the level of significance, symbolized _53_.

53.

The second step requires us to select the desired sample size, _54: symbol?_ .

54.

The third step is to draw a sample of the determined size and calculate a _55: statistic or parameter?_ characterizing the sample, namely the num-

55.

ber of observations of the kind of interest. This count is symbolized __56__
in Appendix K.   56.

Step 4 requires us to go to that appendix and identify the region or
regions of rejection in the relevant __57__ distribution. The sampling   57.
distribution we want is the one for whatever value of P is specified by our
__58__ and for whatever sample size n we chose. Here we will find the   58.
following probabilities: the probability that the count X of the observations of the kind of interest is 0; the probability that this count is 1; the
probability that this count is 2; and so on, up to whatever n is for our
study.

In this distribution of probabilities, we must fix one or two regions of
rejection. If the __59__ hypothesis is nondirectional, there will be two   59.
regions of rejection, one in the lower and one in the upper tail. If the
alternative is directional, we must fix one region of rejection. The one
region goes in the upper tail if the alternative hypothesis says that P is
__60: greater or smaller?__ than the value specified by the null; it goes in the   60.
lower tail if the alternative says that P is __61: greater or smaller?__ than the 61.
value specified by the null.

In all cases, the total area of the region or regions of rejection is
supposed to equal __62: symbol?__ , the level of significance. In practice, it is 62.
usually impossible to achieve a total area of exactly any conventional
level of significance (.05 or .01). Thus the effective level of significance
will differ somewhat from what we chose in Step 1.

To conclude our test, we make the statistical decision, accepting or
rejecting the __63__ hypothesis, depending on whether the outcome that   63.
actually occurred—the __64__ of the observations of the type of interest—   64.
falls in the region of acceptance or the region of rejection.

A Difficulty When the Null Specifies a Probability other than .50 and the Test is Two-Tailed

If the null hypothesis states that $P = .50$, no difficulties arise in the
binomial test, whether the test is one-tailed or two-tailed (except for the
problem that the effective level of significance will usually not be what
we wanted it to be). But if the null states a value other than .50, the
relevant sampling distribution will be __65: symmetrical or skewed?__ . If the   65.
alternative hypothesis states that P is not equal to whatever the null says
—that is, if the test is __66__-tailed—then the problem arises of how to   66.
divide the region of __67__ between the two tails. There is no standard   67.
method for doing this.

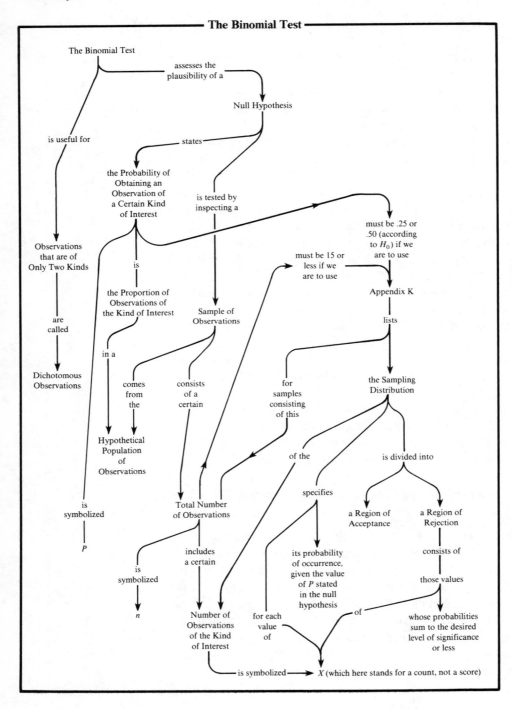

The Binomial Test

Assumptions Underlying the Binomial Distribution

The methods introduced in this chapter assume that the observations are independent of one another and that the probability, 68: symbol? , of   68.
getting an observation of the kind of interest 69: fluctuates or remains   69.
the same? from trial to trial.

SELF-TEST

The symbol P stands for ... A) the probability of one of the two   1.   A   B   C   D
possible results of a single dichotomous observation.   B) the proportion
of observations in a sample that are of a certain kind.   C) "population."
D) "parameter."

The binomial distribution ... A) is actually a family of distributions.   2.   A   B   C   D
B) applies to dichotomous observations.   C) specifies the random sam-
pling distribution of the outcome of a series of trials on each of which we
make a dichotomous observation.   D) is all three of these things.

Use Appendix K of your text to answer this question and the next two. If   3.   A   B   C   D
the probability of an event is .50, which is more likely to happen: it never
occurs in a series of opportunities, or it always occurs in that series?
A) It is more likely never to occur in the series.   B) It is more likely
always to occur in the series.   C) The two outcomes are equally likely.
D) The answer depends on how many trials there are in the series.

If the probability of an event is .25, which is more likely to happen: it   4.   A   B   C   D
never occurs in a series of opportunities, or it always occurs in that
series? A) It is more likely never to occur in the series.   B) It is more
likely always to occur in the series.   C) The two outcomes are equally
likely.   D) The answer depends on how many trials there are in the
series.

Suppose we make an even number of dichotomous observations (2 or 4   5.   A   B   C   D
or 6 or 8, and so on). In which case are we more likely to find that
exactly half of the observations are of one type and half are of the other
type: when the probability of each type is .50, or when the probability of
one type is .25 and the probability of the other type is .75?   A) When
the probability of each type is .50.   B) When the probability of one type
is .25 and the probability of the other is .75.   C) The likelihood is the
same in the two cases.   D) The answer depends on the total number of
observations.

The probability is very close to one-half that a woman who bears a single child in a given birth (as opposed to twins or more) will produce a child who is a girl; let us take it to be exactly one-half. Suppose that on the maternity floor at Fecundity Hospital today, ten women each gave birth to a single child, and nine of them produced a girl. Suppose this outcome puts you to wondering about the probability that ten women who each bear a single child will produce *more than eight* girls. To calculate this probability, ...   A) you cannot use the binomial distribution tabled in Appendix K, because this is not a case of dichotomous observations. B) you cannot use the binomial distribution tabled in Appendix K, because this is not a case of mutually exclusive events.   C) you will have to apply the OR/addition rule.   D) you will have to apply the AND/multiplication rule.

6.   A   B   C   D

In testing a hypothesis by means of the binomial distribution, ...   A) the null hypothesis specifies the probability that a sample of a certain size will contain a certain number of observations of a certain kind.   B) the alternative hypothesis must be nondirectional.   C) the effective level of significance may not be .05 or .01.   D) calculating the statistic that describes the sample is a long, tedious process.

7.   A   B   C   D

In testing a hypothesis by means of the binomial distribution, ...   A) the null hypothesis specifies the probability that one of three or more events will occur.   B) the alternative hypothesis must be directional.   C) sample size must be 20 or more in order for us to get valid results.   D) all observations in our samples must be independent of one another.

8.   A   B   C   D

In testing a hypothesis by means of the binomial distribution, a difficulty will arise if the null hypothesis specifies a probability other than _____ and the alternative hypothesis is _____   A) .25 ... directional.   B) .25 ... nondirectional.   C) .50 ... directional.   D) .50 ... nondirectional.

9.   A   B   C   D

The difficulty cited in Question 9 arises because ...   A) the binomial distribution is skewed when the probability is .50.   B) the binomial distribution is symmetrical when the probability is not .50.   C) statisticians do not agree on how to distribute the region of rejection between two tails of unequal sizes.   D) the binomial distribution does not apply when the probability is not .50.

10.   A   B   C   D

CHAPTER 21

CHI-SQUARE AND FREQUENCY DATA

OBJECTIVES

Ch. 21 presents a technique of inference that is applicable to studies in which we summarize our observations by counting the number that fall into various categories. In studying this chapter, work to learn the following symbols, terms, concepts, principles, and techniques:

The Case of a Single Variable Consisting of C Categories
    Summary of Observations in Terms of the Observed Frequency (f_o) for Each Category
    The Null Hypothesis: a Statement of the Population Proportion (P) in Each Category
    The Alternative Hypothesis
    Expected Frequencies (f_e's)
        f_e for a given category $=(P$ for that category in the population according to $H_0)(n)$
    Chi-Square (χ^2): a Measure of Discrepancy between Observed and Expected Frequencies
        χ^2 cannot be negative
        χ^2 will be zero only if each f_o equals the corresponding f_e
        The larger the difference between an observed and an expected frequency, the larger the value of χ^2
        The value of χ^2 depends on the number of categories
            The Degrees of Freedom (df) for χ^2: $C-1$
        The Random Sampling Distribution of χ^2
            As a Family of Distributions, One for Every Value of the df
            The Shape of a χ^2 Distribution: Positively Skewed
            The region of rejection lies in the upper tail   •   Appendix L: a List of Critical Values
The Case of One Variable with Two Categories
    Dichotomous Observations

The χ^2 Test as an Alternative to the Binomial Test
    The null hypothesis may be a statement about a single proportion
    The alternative hypothesis may be nondirectional or directional
The Case of Two Variables and a Bivariate Frequency Distribution
  Contingency Table with R Rows and C Columns
    Proportionate frequency (p_o) for a cell $= f_o/f_{row}$
    Proportionate frequency for a column (p_{col}) $= f_{col}/n$
      $\Sigma p_{col} = 1.00$
  Independence in a Contingency Table
  The Null Hypothesis
  The Expected Frequencies: f_e for a given cell $= (f_{row})(p_{col}) = (f_{row})(f_{col}/n)$
    In each row and each column, $\Sigma f_e = \Sigma f_o$
  The Alternative Hypothesis
  χ^2
    $df = (R-1)(C-1)$
  Strategy for the Interpretation of a Significant Value of χ^2:
    Inspect the observed proportions row by row or column by column
      The Effect of Sampling Error   •   The Effect of Sample Size on Sampling Error
The 2×2 Contingency Table
  The Shortcut Computational Procedure for χ^2
  The null hypothesis may be a statement about the difference between two proportions
  The Possibility of a One-Tailed Test
The Assumptions of the Chi-Square Test
  Random Sampling
    Statistical generalization is restricted when sampling is not random
  Independence of Observations
    Independence is unlikely when there are more observations than subjects
Special Considerations when Expected Frequencies are Small
  The chi-square test is valid even when f_e's are below 5
  For the test of a single proportion (one variable, two categories), the binomial test is an accurate alternative

Also in studying Ch. 21, work to learn these skills:

Execution of a χ^2 Test for a Study of One Variable, Given Raw Data
  Determination of the f_o's
  Statement of H_0 and H_A
  Choice of α

Computation of f_e's, χ^2, and *df*

Determination of the Region of Rejection via Appendix L

Decision on H_0

Execution of a χ^2 Test for a Study of Two Variables, Given Raw Data

Construction of the Contingency Table

Determination of the f_o's

Interpretation of the Data in Terms of the Proportions Row by Row or Column by Column

Statement of H_0 and H_A

Choice of α

Computation of f_e's, χ^2, and *df*

Determination of the Region of Rejection via Appendix L

Decision on H_0

REVIEW

Overview

A variable is a characteristic that varies. Way back in Ch. 1, on p. 5, the text introduced the distinction between *quant*itative variables and *qual*itative ones. A quantitative variable is a characteristic that varies in quantity. Most of the text has been concerned with quantitative variables of a certain kind: those on which the observations consist of numbers (scores) that represent the quantities of the varying characteristic. An example is birthweight, for observations on this variable consist of scores representing the quantity of weight. A quantitative variable of this kind is said to be __1__ scaled.

1.

Most of the techniques of description and inference presented in the text are appropriate only for quantitative variables that are numerically scaled—appropriate, that is, only for observations that are scores. It is scores that are described by the grouped frequency distribution of Ch. 2; by the histogram, the frequency polygon, and the cumulative percentage curve of Ch. 3; by the mean and standard deviation of Chs. 4 and 5; by the normal-curve model and the z scores of Ch. 6; and by the Pearson correlation coefficient of Ch. 7. It is also scores that enter into the inferential techniques of Chs. 8 and 10–19, the techniques employing the regression equation and the t and F statistics.

In Ch. 21, the text returns to the other kind of variable, the __2__ kind. In recording observations on this kind of variable, we do not assign scores to our subjects or do anything else that indicates the quantity of a

2.

varying characteristic. Instead, we classify each subject into a category. If we were studying consumer preference among various brands of cola drink, for example, we would classify a subject into a category such as "Preferred Coca Cola" or "Preferred Pepsi Cola." We summarize qualitative observations not by calculating a mean and a standard deviation, but by simply counting the subjects classified into each category. Such a count is called a 3 .

3.

(Ch. 20. also dealt with observations of the qualitative kind. The binomial test introduced there applies to a qualitative variable that consists of only two categories, such as sex, which can be only male or female.)

The technique of inference presented in Ch. 21 (like the technique of Ch. 20) can be used for a quantitative variable that is numerically scaled if we first group the scores into class intervals. (See pp. 16–17 to review this matter.) We must then treat each class interval as a(n) 4 .

4.

The Case of One Variable

In the simplest case of qualitative observations, there is just one variable of interest, and it consists of two or more categories. We summarize the data by counting the subjects classified into each category, and we call such counts frequencies, as noted above. Look at the *top* row of Table 21.1 on p. 369 to see the frequencies that summarize a hypothetical study of consumer preference in which there is just one variable of interest: choice among four brands of cola. In this case, the variable consists of 5: how many? categories. (The latter portion of this chapter deals with the case in which there are two qualitative variables under study, and we simultaneously classify a subject into one category of the first variable *and* one category of the second variable.)

5.

The Null Hypothesis and the Expected Frequencies

In most cases, the observations on hand will be just a sample, not the entire set of observations of interest, and we will wish to draw a conclusion about this population of observations. Specifically, we will want to know the 6: numbers or proportions? of the observations in the population that fall into the various categories. To draw a conclusion about these proportions, we can apply the logic of hypothesis testing.

6.

As usual, the hypothesis to be tested will be an assertion about the 7: sample or population? , and it will be called the 8 hypothesis, symbolized 9 . Here, the hypothesis will specify what proportion of the population of observations falls into each possible category of the

7.

8.

9.

variable under study. The null _10: must or need not?_ specify equal proportions.   10.

To assess the plausibility of the null hypothesis, we need to calculate some quantities called expected frequencies, symbolized _11_ ("ef sub   11.
ee"). There is one expected frequency for each category, and it is the mean of the frequencies that would obtain for that category over an infinitely long series of replications of our study *if the null hypothesis were*
12: true or false? .   12.

The calculation of an expected frequency is simple. For the given category, multiply these two numbers: the proportion specified by the _13_ hypothesis for that category and the _14_ . The bottom row of the   13.
table on p. 369 shows the expected frequencies for the study of consumer   14.
preference among cola drinks.

Now that we are talking about expected frequencies, we need a term to designate the other kind of frequency, the kind that is the count of the observations that actually occurred. The other kind is called a(n) _15_   15.
frequency, symbolized _16_ ("ef sub oh").   16.

There is a way to check the computation of the expected frequencies. The sum over all categories should equal the sum of all the observed frequencies—and the sum is the sample size, n. In symbols, $\Sigma f_e =$ _17_ .   17.

Note that an expected frequency may have a fractional part. (For example, if that study of consumer preference had employed a total of 50 subjects, each expected frequency would have been $(.25)(50) = 12.5$.) An *observed* frequency _18: can or cannot?_ have a fractional part, because it   18.
is a count, a count of the observations falling in a certain _19_ , and there   19.
is no such thing as a fraction of an observation. But an expected frequency is not a count. It is a mean, as noted above, and as such it can properly include a fractional part.

Note also that the expected frequencies for the various categories are those that make the sample exactly resemble the population as the null hypothesis envisions the population—except that there are _20: more or   20.
fewer?_ cases in the population. In the study of taste preference, the expected frequencies show an even distribution, with _21: what   21.
fraction?_ of the cases in each category, and the null hypothesis states that the population shows this same even distribution.

The Chi-Square Statistic as a Measure of Discrepancy between Observed and Expected Frequencies

If the null hypothesis is _22: true or false?_ , each observed frequency will   22.
probably be about the same size as the corresponding expected frequency. But if the null is _23: true or false?_ , the observed frequencies will proba-   23.

bly differ somewhat from their counterparts among the expected frequencies. Thus a set of observed frequencies similar to the expected ones will suggest that we should __24__ the null hypothesis, while a set of observed frequencies considerably different from the expected ones will suggest that we should __25__ the null.

Clearly, we need a device for measuring the discrepancy between the observed and the expected frequencies. Such a device is the statistic called chi-square, symbolized __26__. To calculate chi-square, we must go through these simple steps for each category:

　　1. Subtract the __27__ frequency from the __28__ frequency. In symbols, calculate __29__.

　　2. Square the difference. In symbols, calculate __30__.

　　3. Divide the difference by the __31__ frequency. In symbols, calculate __32__. The final step is to add the results obtained in Step 3 for each category. That is:

　　4. Calculate __33: symbols?__.

Chi-square cannot be negative. Can it be zero? Yes, but only if each observed frequency is exactly equal to the corresponding __34__. This will be unlikely if the null hypothesis is false, of course, but it will also be unlikely even if the null hypothesis is true.

What affects the size of chi-square? Other things equal, the larger the difference between the observed and the corresponding expected frequencies, the __35: larger or smaller?__ the value of chi-square. It __36: matters or does not matter?__ whether an observed frequency is larger or smaller than the corresponding expected frequency; that is, the direction of the difference is irrelevant. All that counts, so far as chi-square is concerned, is the size of this difference. Thus chi-square is sensitive to the difference between the observed and the expected frequency for each category, and it is this sensitivity that makes it a measure of the discrepancy between the actual results in the sample and those we would expect if the null hypothesis were __37: true or false?__.

There is another thing that affects the size of the chi-square statistic, though, and that is the __38__ of categories. For each category, we calculate a difference between the observed and the expected frequency, square the difference, and divide the square by the __39__ frequency. We then sum the resulting quantities over all the categories. The sum *is* chi-square. Obviously, the more categories there are, the more opportunities there are for the chi-square statistic to become large. In judging the size of a chi-square value, then, we must take this matter into account.

We do this by considering the degrees of __40__ associated with the chi-square statistic, for they are closely related to the number of cate-

24.

25.

26.

27.
28.
29.
30.
31.
32.
33.

34.

35.
36.

37.

38.

39.

40.

gories. If C stands for the number of categories, the degrees of freedom are given by the formula $df = \underline{\;41\;}$.

41.

(For the t statistic and for the within-groups and within-cells variance estimates that contribute to the F statistic, the degrees of freedom depend on the size of the $\underline{\;42\;}$. But for chi-square, sample size has nothing to do with the degrees of freedom.)

42.

The Random Sampling Distribution of Chi-Square

The logic underlying a test of a hypothesis by means of a chi-square value is $\underline{\;43:\text{ the same as or different from?}\;}$ the logic underlying any other statistical test of a hypothesis. We calculate a number on the basis of a sample; such a number is a statistic in the technical sense of that word. We ask what values of this statistic would be likely to occur if the null hypothesis were $\underline{\;44:\text{ true or false?}\;}$. If the value that we obtained is one that would be likely if the null were true, we will $\underline{\;45\;}$ the null—though we would not have proved it be true, of course. (See pp. 200–201 to review this matter.) If the value we obtain for the statistic is one that would be rare if the null were true, we will $\underline{\;46\;}$ the null, concluding that it is not plausible. And exactly what we do mean by "likely" and "rare"? That's specified by our level of $\underline{\;47\;}$, symbolized α. If we set α at .05, we are deciding that we will regard as likely a number whose probability is greater than $\underline{\;48\;}$, and as rare a number whose probability is .05 or less.

43.

44.
45.

46.

47.

48.

How can we tell which values of our statistic would be likely and which would be rare if the null were true? The answer comes from the $\underline{\;49\;}$ distribution of the statistic. This distribution consists of the values of the statistic that would occur in an infinitely long series of replications of our study if the null hypothesis were $\underline{\;50:\text{ true or false?}\;}$. Statisticians have determined what the characteristics of this distribution would be for the t ratio, for the F ratio, and for many other statistics.

49.

50.

In the case of chi-square, as for t and F, there is not just one sampling distribution but a family of them, one for every value of the $\underline{\;51\;}$. Every member of the family has a shape that is $\underline{\;52\;}$, as shown in the figure on p. 371.

51.
52.

The Chi-Square Test

Now we can put all of this information together into a test of a null hypothesis. The null, to repeat, states what $\underline{\;53\;}$ of the $\underline{\;54:\text{ population or sample?}\;}$ falls into each category of the one variable that is under study. The alternative hypothesis says simply that the null is $\underline{\;55\;}$; it thus

53.
54.

55.

covers a great many possibilities. (An exception—a case in which the alternative can be more specific—is described below.) The level of significance is whatever we choose; it is conventional to set it at __56__ .   56.

After determining the sample size we desire, we draw a sample of that size, ideally __57: how?__ , from the population of interest. We calculate the   57. observed and the expected frequency for each category and go on to figure chi-square.

We must then identify the region of rejection in the relevant sampling distribution. This is the sampling distribution of chi-square for the degrees of __58__ that apply to our study. Do we want the region of   58. rejection to appear in the lower tail, in the upper tail, or divided between the two? We want there to be one region, and it should go in the __59__ tail.   59. Why? Because it is only the __60: large or small?__ values of chi-square that   60. are likely to occur when the null hypothesis is false. The figure on p. 371 provides an illustration. So we look up the appropriate critical value of chi-square in Appendix L.

In the last step, we accept the null hypothesis if the obtained value of chi-square is __61__ the critical value, and if the obtained equals or exceeds   61. the critical value, we __62__ the null.   62.

The Case of Two Categories

If our variable consists of just two categories, the observations on the variable are called __63__ . Such observations were the subject of Ch. 20,   63. and that chapter introduced the binomial test as a way to assess the plausibility of a null hypothesis specifying the probability that one of the two possible observations will occur.

The binomial test works well with small samples, but it is cumbersome to apply when sample size is large. (And the relevant table in your text, Appendix K, goes up to a sample size of only 15.) Fortunately, for large samples, the __64__ statistic provides a close approximation to the exact   64. results that would come from the binomial distribution.

To use the chi-square statistic, we proceed in the usual way. The null hypothesis can be conceived either as a statement of the probability that one of the two possible observations will occur or as a statement of the __65__ of all observations that are of this kind.   65.

The degrees of freedom for a chi-square derived from two categories are __66__ , and in this case and only in this case, it is possible to state an   66. alternative hypothesis that is __67__ . There are two types of directional   67. alternative: one says that the probability or the population proportion is less than what the null specifies, and the other says that the probability or the population proportion is greater than what the null specifies. If we conduct a one-tailed test, the critical value of chi-square cannot be found in Appendix L. For the 5% significance level it is 2.71, and for the 1% level it is 5.41. We should reject the null hypothesis only if the sample is the kind that the _68: null or alternative?_ hypothesis envisions.

68.

Application of Chi-Square to a Study of Two Variables

The chi-square statistic is also applicable to studies in which the researcher makes observations on two qualitative variables simultaneously. Here, the researcher classifies each subject into one category of one variable *and*, at the same time, into one category of the other variable. Data of this kind are called bivariate. In the hypothetical study diagrammed in Table 21.2 on p. 375, 200 students were classified into one of two categories that make up a variable that we could call "Academic Division to which Student Belongs." Each category of this variable is represented by a _69: row, column, or box?_ in the table. The 200 students were also classified into one of three categories of a second variable that is the student's preference among the possible roles that the student body could play in college governance. The categories of this variable are represented by the _70: rows, columns, or boxes?_ of the table.

69.

70.

To summarize data of this kind, we need only count the observations that fall in each *combination* of a category from one variable and a category from the other. Such combinations are represented by the _71: rows, columns, or boxes?_ of the table on p. 375. The boxes are called cells. The count of the observations in a given cell is a frequency; the frequencies that actually occurred are termed _72_ frequencies.

71.

72.

Independence and Dependence

A table like the one on p. 375 is called a _73_ table, and it permits us to tell whether one variable is contingent upon the other for the sample on hand. This is the issue of whether the two variables are related to each other or whether they are independent of one another. (The question is similar to that of whether there is a correlation between two numerically scaled variables for which the observations consist of scores; see Ch. 7 for a review.)

73.

To determine whether the two variables are independent or dependent in the sample of observations shown in a _74_ table, we must work either

74.

row by row or column by column. We will get the same answer to the question of independence whichever way we do it. Let us do it by rows. In the first row, we find the total number of cases (for just that row). The symbol for this quantity is __75__. We then find the proportion of this total that falls into each cell along the row. If we use the symbol f_o for the frequency observed in a given cell and the symbol p_o for the corresponding proportion relative to the row total, the formula for calculating p_o is __76__. We must find p_o for each cell in the first row. We then repeat for the second row, which may have a different total, and repeat again for any other rows. The table on p. 376 shows the calculations for the study of student opinions.

75.

76.

What do the resulting proportions tell us? If the set of proportions for the cells is exactly the same along each row, there is *no* relationship between the two variables for the sample on hand; that is, the two variables are __77: dependent or independent?__ in the sample. If there is any variation in the set of proportions for the cells from one row to another, then there is some relationship between the two variables for the sample on hand; that is, the two variables are __78: dependent or independent?__ in the sample. (Another way to say this is that one variable is contingent on the other.)

77.

78.

As an example, consider the hypothetical study in which 200 students responded to an interview or a questionnaire. (The sociological terminology for them is *respondents.*) The results are shown in the tables on pp. 375 and 376. If the variables under study were *in*dependent (if there were *no* relationship) for this sample, the role a respondent prefers for the student body would have nothing to do with whether the respondent is affiliated with the Division of Science and Technology or the Division of Business. A certain proportion of the respondents would prefer that students play an advisory role; a certain proportion would prefer that some students be nonvoting members of governance committees; a certain proportion would prefer that some students be voting members— and these proportions would be the same for respondents from Science and Technology as they are for respondents from Business. That is, the category (the row) into which a respondent falls on the variable Academic Division would make no difference for the likelihood that she or he falls into one category or another (one column or another) on the other variable.

The results shown in the tables indicate, however, that the two variables are not independent in the sample. They are dependent (related). How can we tell? The proportions in the top row, those for the Science-Technology respondents, don't match the proportions in the bottom row for the Business respondents. The pattern in the data is not simple, but it appears that the Business respondents were more conservative on the issue put to them. That is, a Business student in the sample was more likely to advocate only an advisory role for the student body; _79: what proportion?_ of them did so, but only _80: what proportion?_ of the Science-Technology respondents did. In contrast, over half of the Science-Technology respondents (.60) advocated the more activist position that students should serve as _81: voting or nonvoting?_ members of governance committees, while only .35 of the Business respondents took this stand.

79.
80.

81.

Testing a Hypothesis about the Relationship between the Variables in the Population

Even if two variables are independent in a population, a sample from that population can still show a relationship between them. Thus we cannot automatically conclude that there is a relationship in a population when we find one in a sample from that population. To draw a proper conclusion on the basis of a sample, we must employ the logic of hypothesis testing and the chi-square statistic.

The null hypothesis states that the variables are _82_ in the population. This means that the distribution of proportionate frequencies _83: varies among the or is the same for all?_ rows. (We could also think the problem through in terms of columns and get the same results.) The alternative hypothesis says simply that the null is false, and there are many ways in which it could be false. (An exception occurs if there are only _84_ categories for both variables, as described below.)

82.
83.

84.

To test the null using chi-square, we must find the expected frequency for each cell in the contingency table. The procedure is this: Find the total number of observations in the column in which the cell is located. This quantity is symbolized _85_ . Convert this to a proportion by dividing it by the grand total, symbolized _86_ . The result is the proportion symbolized p_{col}. (That is, $p_{col} = f_{col}/n$.) Then multiply the _87_ for the row in which the cell is located by the proportion for its column.

85.
86.
87.

There are two ways to check the computations. Across the several columns, the column proportions, symbolized p_{col}, should total to _88_ . And in each row and each column, the total of the expected frequencies should equal the total of the _89_ .

88.

89.

We can now calculate the chi-square statistic in the same way we would for a study in which there is just one variable.

The only difference between the rest of the procedure for a study of
two variables and that for a study of one variable lies in the computation
of the degrees of freedom. When there are two variables, the degrees of
freedom are calculated as one less than the number of rows
90: plus or times? one less than the number of _91_. If R stands for the
number of rows and C for the number of columns, the formula is:
$df =$ _92_.

90.
91.
92.

A convenient summary of the chi-square test of independence with all
the steps shown explicitly appears on p. 379 of the text.

When one or both of the variables consist of several categories, a
significant value of chi-square may not be very informative. How exactly
are the two variables related in the population, we will probably have to
ask. Some clues can come from a close inspection of the observed
proportions row by row (or column by column), as in the table on p. 378.
But we must remember that these proportions are only estimates of the
corresponding values characterizing the population; they are subject to
_93_ error. If the sample is large, though, the error will probably be
relatively _94: large or small?_.

93.

94.

The 2×2 Contingency Table

When each variable consists of just two categories, the contingency table
is a 2×2 ("two by two") affair, and three special considerations arise.

First, we can calculate chi-square by means of a formula that bypasses
the expected frequencies and saves computational labor. The formula
appears on p. 380.

Second, we can conceptualize the null hypothesis in the usual way, as
the assertion that the two variables are _95_ in the population, _or_ in a
different way. We can envision two populations, one for each row. The
null would then state that there is no difference between the proportion
of one population that falls into a certain category of the other variable
and the corresponding proportion for the other population. In the
example shown in the table on p. 381, we envision a population of
observations on men, from which we have the sample of _96_ tabulated in
the _97: top or bottom?_ row, and a population of observations on women,
from which we have the sample of 90 shown in the bottom row. As a null
hypothesis, we assert that in the populations, the proportion of men who
fall into the left-hand column (who aspire only to a master's degree)
equals the proportion of women who fall into that column. (We could
also think of each column as a separate population and proceed in an
analogous fashion.)

95.

96.
97.

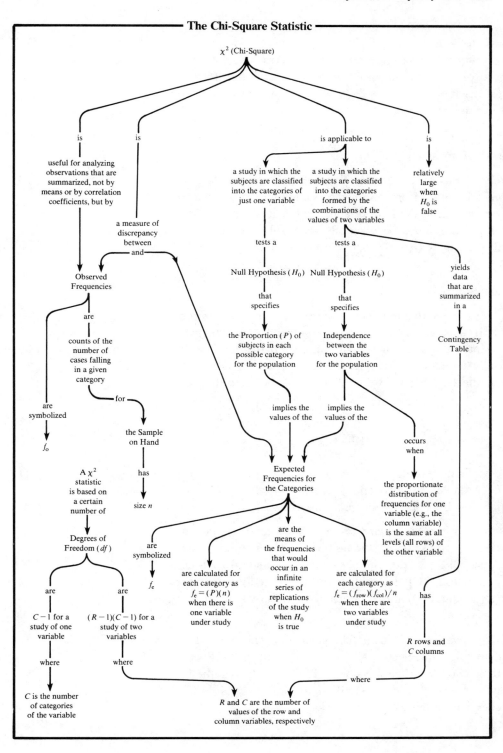

The Chi-Square Statistic

χ^2 (Chi-Square)

is

useful for analyzing observations that are summarized, not by means or by correlation coefficients, but by

is

a measure of discrepancy between —— and——

is applicable to

a study in which the subjects are classified into the categories of just one variable

a study in which the subjects are classified into the categories formed by the combinations of the values of two variables

is

relatively large when H_0 is false

Observed Frequencies

are

counts of the number of cases falling in a given category

are symbolized

f_o

for

the Sample on Hand

has

size n

tests a

Null Hypothesis (H_0)

that specifies

the Proportion (P) of subjects in each possible category for the population

tests a

Null Hypothesis (H_0)

that specifies

Independence between the two variables for the population

yields data that are summarized in a

Contingency Table

A χ^2 statistic is based on a certain number of

Degrees of Freedom (df)

are

$C-1$ for a study of one variable

are

$(R-1)(C-1)$ for a study of two variables

are symbolized

f_e

implies the values of the

implies the values of the

Expected Frequencies for the Categories

occurs when

the proportionate distribution of frequencies for one variable (e.g., the column variable) is the same at all levels (all rows) of the other variable

are calculated for each category as $f_e = (P)(n)$ when there is one variable under study

are the means of the frequencies that would occur in an infinite series of replications of the study when H_0 is true

are calculated for each category as $f_e = (f_{row})(f_{col})/n$ when there are two variables under study

has

R rows and C columns

where

where

where

C is the number of categories of the variable

R and C are the number of values of the row and column variables, respectively

Third, since the degrees of freedom for the chi-square statistic here are
_98_, it is possible to do a _99_-tailed test. Our alternative hypothesis can
specify the direction of the relationship between the two variables (if we
conceptualize the problem in the usual way), or it can specify which
proportion is greater than which (if we conceptualize the problem in the
way described just above).

98.
99.

Assumptions of the Chi-Square Test

The chi-square test rests on two assumptions: sampling was _100_, and
the observations are _101_ of one another. If sampling was not random,
we are sharply restricted in the kind of statistical generalization that we
can draw. If there are more observations than subjects, the second
assumption will very likely be _102: right or wrong?_.

100.
101.

102.

The Role of Sample Size

Sample size plays the same role in the chi-square test that it does in other
statistical tests. If sample size is _103: large or small?_, a statistically
significant value for chi-square can easily occur even if the null hypothesis is only trivially false. But if sample size is _104: large or small?_, the
null can be dramatically false, yet we may be unable to detect this state
of affairs.

103.

104.

SELF-TEST

The chi-square statistic permits us to test a null hypothesis about ...
A) at most one variable.   B) two and only two variables.   C) either
one or two variables.   D) one, two, or three variables.

1.   A   B   C   D

The null hypothesis for a chi-square test in which there is only one
variable of interest specifies ...   A) the number of observations in each
category for the sample.   B) the number of observations in each category for the population.   C) the proportion of observations in each
category for the sample.   D) the proportion of observations in each
category for the population.

2.   A   B   C   D

For a study of one variable, a researcher calculated chi-square to be
-6.61. What does this number tell us?   A) Most or all of the f_o values
are less than their corresponding f_e values.   B) The null hypothesis is
almost certainly true.   C) The null hypothesis is almost certainly false.
D) The researcher made a mistake in calculating chi-square.

3.   A   B   C   D

The null hypothesis for a chi-square test in a study of genetics says that among all the children that two particular parents could produce, one-quarter will be severely retarded, one-half will be mildly retarded, and one-quarter will develop normally. Sample size is 6 children from these two parents. What is f_e for the category "severely retarded"?   A) One-quarter.   B) .25   C) 1.5   D) 3

4.   A   B   C   D

What are the *df* for the chi-square statistic of Question 4?   A) 1   B) 2   C) 3   D) There is not enough information to tell.

5.   A   B   C   D

What will be the sum of the expected frequencies for the chi-square statistic of Question 4?   A) 1.00   B) 2.00   C) 3.00   D) 6.00

6.   A   B   C   D

What is the shape of the random sampling distribution of chi-square?   A) Skewed left.   B) Symmetrical.   C) Skewed right.   D) Rectangular.

7.   A   B   C   D

The alternative hypothesis for a chi-square test may be directional when ...   A) there is only one variable under study.   B) there are more than two categories.   C) the level of significance is below .01.   D) there is one degree of freedom.

8.   A   B   C   D

A contingency table displays the results of a study of ...   A) one variable.   B) two variables.   C) either one or two variables.   D) more than two variables.

9.   A   B   C   D

If we obtain a statistically significant value of chi-square for a contingency table, we should conclude that in the population from which we sampled, the variables under study are probably ...   A) independent of one another.   B) related to each other.   C) distributed in the same way.   D) distributed in different ways.

10.   A   B   C   D

CHAPTER 22

SOME (ALMOST) DISTRIBUTION FREE TESTS

OBJECTIVES

Ch. 22 presents three techniques of inference that are sometimes useful when sample sizes are small. In studying the chapter, work to understand the following symbols, terms, concepts, principles, and techniques:

The Sign Test
    As an Alternative to the t-Test for Dependent Samples
    Assignment of $+$ or $-$ to Each Pair of Scores
       Dealing with a Pair of Identical Scores
    The Null Hypothesis
    The Alternative Hypothesis
    Conduct of the Sign Test via χ^2
       Observed Frequencies (f_o's)
       Expected Frequencies (f_e's)
       χ^2
       Appendix L: a List of Critical Values of χ^2
    Conduct of the Sign Test via the Binomial Distribution
    Assumptions of the Sign Test

Also in studying Ch. 22, work to learn these skills:

Rank-Ordering Scores
Execution of a Mann-Whitney Test, Given Scores
    Statement of H_0 and H_A •   Choice of Level of Significance   •   Calcu-
    lation of ΣR_X •   Determination of the Critical Value(s)   •   Decision on
    H_0 •   Calculation of z when Sample Size is Large   •   Evaluation of z
Execution of a Kruskal-Wallis Test, Given Scores
    Statement of H_0 and H_A •   Choice of Level of Significance   •   Calcula-
    tion of ΣR_1, ΣR_2, ...   •   Calculation of H •   Determination of the
    Critical Value of χ^2 •   Decision on H_0
Execution of a Sign Test, Given Scores
    Statement of H_0 and H_A •   Choice of Level of Significance   •   Assign-
    ment of Pairs of Scores to the Category $+$ or the Category$-$ •   Calculation
    of χ^2 •   Determination of the Critical Value of χ^2 •   Decision on H_0
      •   Execution of the Binomial Test for Sample Size below 10

REVIEW

Overview

The text has now presented three techniques for testing a hypothesis
about the means of two or more populations. For a study in which the
samples are independent and there are just two populations, the text
described a test based on the __1__ statistic. For a study in which the
samples are independent and there are three or more populations of
interest, the text described a test based on the __2__ statistic. And for a

1.

2.

study in which the samples are dependent and there are two populations
of interest, the text described a test again based on the __3__ statistic.   3.

   These techniques make assumptions about the distributions of the
scores in the populations. All three assume that every population of
interest has a shape that is __4__. The two tests for independent samples   4.
also assume that the populations all have the same __5__. When sample   5.
sizes are __6: small or large?__, these tests will yield accurate results even if   6.
their underlying assumptions are wrong, even substantially wrong. But a
problem can arise when the assumptions about the distributions in the
populations are materially violated and sample sizes are small.

   Ch. 22 presents alternative procedures that make less restrictive assumptions about the __7: sample or population?__ distributions. They are   7.
sometimes called distribution-free methods, but they are not completely
free of underlying assumptions about the distributions of the scores in
the populations; their assumptions are just less likely to be wrong. (Thus
the title of the chapter refers to tests that are "*almost* distribution free.")
These alternative tests are also called __8__ methods.   8.

   Although the alternatives described in this chapter are very useful
when the assumptions of the *t*- and *F*-tests are wrong and samples are
small, when those assumptions are correct (and regardless of sample
size), the alternatives are less sensitive. The alternatives, that is, are less
likely to detect a difference between the populations by yielding a result
that is statistically __9: "significant" or "nonsignificant"?__. In technical terms,   9.
the tests are less __10__.   10.

   As a measure of the central tendency (location) of a distribution, the *t*-
and *F*-tests employ the __11__. The alternatives are less specific. They do   11.
not test a null hypothesis that says that the populations of interest all
have the same mean or are otherwise the same in any other measure of
__12__. Instead, the alternative techniques test a null hypothesis asserting   12.
that the populations are identical in all respects. This is mildly annoying,
because one may get significant results and thus __13: accept or reject?__ the   13.
null hypothesis due to some unknown combination of differences in the
central tendencies, the variabilities, and the shapes of the populations.
But if the distributions are not greatly different in __14: which two__   14.
__characteristics?__, these tests are good at determining whether the populations are identical in their central tendency (location).

Placing Scores in Rank Order

The first two tests discussed in this chapter require us to place the scores
in our samples in order of magnitude from lowest to highest; that is, we
must translate each score into a __15__. A problem that frequently arises   15.

here is what to do with identical scores, for they lead to ties in rank. For example, the two smallest scores should receive the ranks __16: what and what?__ . If there are two scores tied for the smallest value, we have a problem, for they cannot both be given rank 1. The solution is to assign to each of the scores tied for a given rank the __17__ of the ranks that are available to those scores. In the example, we would assign to each of the two scores tied for first place the mean of the ranks 1 and 2, which is __18__. This procedure will normally be satisfactory unless perhaps as many as __19: what fraction?__ of all the scores are involved in ties.

It is sometimes easy to make a mistake in assigning ranks. We can check our work by determining whether the sum of all the ranks equals $n(n + 1)/2$, where n is the number of __20__. This should be the sum __21: only when there are no or even if there are some?__ ties.

16.

17.

18.
19.

20.
21.

An Alternative to the *t*-Test for Independent Samples

As an alternative to the *t*-test for independent samples, the text presents the __22__. As noted above, the null hypothesis asserts that the two populations are identical in __23: central tendency or in all ways?__ . Nevertheless, if the two populations are even moderately similar in variability and in shape, this is an excellent test of whether they are identical in __24__. The best measure of central tendency to use in connection with the test is the __25__.

The alternative hypothesis may be nondirectional or directional. If it is __26__, it says that the two populations are different, but it does not say which has the higher scores. If the alternative is directional, it takes one of two forms. One says that one population has the larger scores; the other says that the other population has the larger scores.

To conduct the test, label the two samples X and Y. If they are of unequal sizes, in order to use the table of critical values in Appendix M, the sample with fewer cases must be called __27__. Combine all scores into one big distribution. Assign the rank of 1 to the __28: highest or lowest?__ score; assign the rank 2 to the next __29: highest or lowest?__ ; and so on, until all scores are ranked. Then find the sum of the ranks of all the scores in the distribution labeled __30__. The sum is symbolized __31__.

Appendix M provides the critical values for the statistic ΣR_X. The critical value or values that apply to a given study depend on the level of significance we choose and on whether the alternative hypothesis is __32__. Instructions for the use of the table appear on p. 392.

When sample sizes are too large to use Appendix M, we can convert ΣR_X to a statistic appropriately called __33__. The shape of the random

22.
23.

24.

25.

26.

27.
28.
29.

30.
· 31.

32.

33.

sampling distribution of this statistic, when the null hypothesis is true, is approximately __34__ . Thus we can use the table of areas under the normal curve (Appendix F) to determine the critical value or values of z.

34.

The assumptions underlying the Mann-Whitney test are __35__ sampling and no ties in ranks. A moderate number of tied ranks __36: will or will not?__ invalidate the test.

35.
36.

An Alternative to the *F*-Test for Independent Samples

As an alternative to the F-test in the one-way analysis of variance, this chapter presents the __37__ test. Just as the F-test is an extension of the __38__ test from two independent samples to more than two, so the __39__ test is an extension of the __40__ test from two independent samples to more than two.

37.
38.
39.
40.

Again, the null says that the several populations are entirely identical, but under ordinary circumstances, the Kruskal-Wallis is a good test of whether their __41__ are the same. Again, the best measure of central tendency is the __42__ , but we might also use the mean if the shapes are not badly skewed. The alternative hypothesis says simply that the null is false; it could be false in __43: many ways or only one way?__ .

41.
42.

43.

To conduct the Kruskal-Wallis test, we must again combine the scores in the samples into one large group and rank them from 1 (the __44: highest or lowest?__ on up. Then for each sample we must find the sum of the ranks of the scores in the sample. The sum for the first sample (and it doesn't matter which we call the first) is symbolized __45__ ; the sum for the second sample is symbolized __46__ ; and so on. Using the formula on p. 394, we then calculate the statistic called H.

44.

45.
46.

When the null hypothesis is true, values of H tend to be small; when the null is false, values of H tend to be large. To evaluate the size of H, we can use the random sampling distribution of __47__ , which is shown in Appendix L. The degrees of freedom are $k - 1$, where k is the number of __48__ . The region of rejection always appears in the __49: lower or upper?__ tail of the distribution.

47.

48.
49.

If there are three samples, the chi-square distribution will be valid for use in the Kruskal-Wallis test if there are at least __50__ scores per sample. When there are more than three samples, some can have as few as two or three scores.

50.

The assumptions underlying the Kruskal-Wallis test are the same as those underlying the __51__ test: __52__ sampling and no __53__ . Like the other test, the Kruskal-Wallis remains valid unless there are many ties.

51.
52.
53.

An Alternative to the *t*-Test for Dependent Samples

As an alternative to the *t*-test for dependent samples, the text offers the __54__ test. Since the samples are dependent, the scores come in pairs, each consisting of a score from the first sample and a score from the second. Look at the table on p. 396 for an example. To conduct the test, we inspect each pair of scores, and we assign to it a plus (+) or a minus (−). A pair gets a plus if the score from the second sample is larger; it gets a minus if the score from the first sample is larger. Thus the sign, plus or minus, indicates the direction of the difference between the two scores; that is, it indicates which score is larger.

54.

If some of the differences are zero, we should simply ignore them and reduce the value of __55__ accordingly.

55.

The null hypothesis says that in the __56: sample or population?__ , the proportion of pluses and the proportion of minuses are equal (.50 for each). The alternative hypothesis may be nondirectional or directional. If it is nondirectional, it asserts that the proportions are __57__ without saying exactly how. If it is directional, it takes either of two forms. In one form, it says that the proportion of pluses is greater; in the other form, it says that the proportion of __58__ is greater.

56.

57.

58.

To test the __59__ hypothesis, we can employ the __60__ statistic. We work with two categories, plus and minus. The count of the cases in each category is a(n) __61__ , symbolized f_o. The expected frequency for the pluses will be __62: what fraction?__ of the sample size, and the expected frequency for the minuses will be the same. (The calculation of an expected frequency is an application of the rule on p. 368: for a given category, multiply the __63__ of cases that are supposed to fall into that category in the population, according to the null hypothesis, by the sample size.)

59.

60.

61.

62.

63.

Having determined each observed frequency, f_o, and the corresponding expected frequency, __64: symbol?__ , we go on to figure the value of __65__ .

64.

The degrees of freedom for the chi-square statistic when we use it in the sign test are __66__ . If the alternative hypothesis is __67: directional or nondirectional?__ , we can obtain the critical value from the table in Appendix L. If the alternative is __68__ , we must use the critical value specified on p. 374: 2.71 for the .05 level of signficance or 5.41 for the .01 level.

65.

66.

67.

68.

Conducting the sign test by means of the chi-square statistic will give reasonably accurate results for 10 or more pairs of scores. If there are fewer than 10, we should apply the __69__ test.

69.

Two assumptions underlie the sign test: sampling is __70__ , and no difference is exactly zero. The test will still give valid results if we ignore

70.

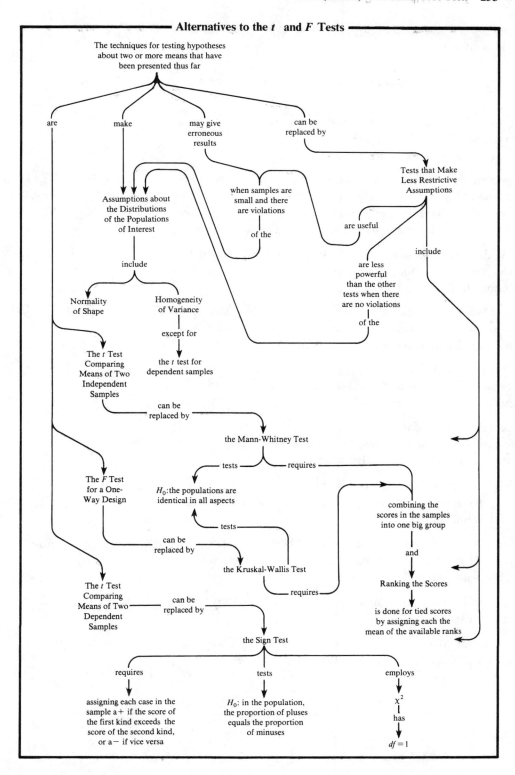

Alternatives to the _t_ and _F_ Tests

The techniques for testing hypotheses about two or more means that have been presented thus far

are

make

may give erroneous results

can be replaced by

Tests that Make Less Restrictive Assumptions

Assumptions about the Distributions of the Populations of Interest

when samples are small and there are violations

of the

are useful

include

include

are less powerful than the other tests when there are no violations

of the

Normality of Shape

Homogeneity of Variance

except for

the _t_ test for dependent samples

The _t_ Test Comparing Means of Two Independent Samples

can be replaced by

the Mann-Whitney Test

tests

requires

combining the scores in the samples into one big group

The _F_ Test for a One-Way Design

H_0:the populations are identical in all aspects

and

tests

Ranking the Scores

can be replaced by

the Kruskal-Wallis Test

is done for tied scores by assigning each the mean of the available ranks

requires

The _t_ Test Comparing Means of Two Dependent Samples

can be replaced by

the Sign Test

requires

tests

employs

assigning each case in the sample a + if the score of the first kind exceeds the score of the second kind, or a − if vice versa

H_0: in the population, the proportion of pluses equals the proportion of minuses

χ^2

has

$df = 1$

any differences of zero and reduce __71__ accordingly, provided the num-   71.
ber of zeroes is small.

SELF-TEST

Which of the following tests assumes that the populations of interest all   1.   A   B   C   D
have a normal shape?   A) The Kruskal-Wallis test.   B) The Mann-
Whitney test.   C) The sign test.   D) The t-test.

When the assumptions underlying the test are correct, which test is more   2.   A   B   C   D
likely to reject the null hypothesis if the null is false?   A) The t-test for
independent samples.   B) The Mann-Whitney test.   C) The two tests
are equal in this regard.   D) The answer depends on the sample size.

The Kruskal-Wallis test is an extension of the ...   A) t-test for indepen-   3.   A   B   C   D
dent samples.   B) t-test for dependent samples.   C) sign test.
D) Mann-Whitney test.

The null hypothesis for the Mann-Whitney test asserts that the popula-   4.   A   B   C   D
tions are identical in ...   A) their central tendency.   B) their variability.
C) their shape.   D) all characteristics.

The null hypothesis for the Kruskal-Wallis test asserts that the popula-   5.   A   B   C   D
tions are identical in ...   A) their central tendency.   B) their variability.
C) their shape.   D) all characteristics.

A sample of scores consists of the values 84, 85, 87, 88, 88, 88, 88, 89, and   6.   A   B   C   D
90. When we rank-order these scores, the first score of 88 on the list
should receive the rank of ...   A) 1.   B) 4.   C) 5.5.   D) 22.

In the Mann-Whitney test, the alternative hypothesis ...   A) must be   7.   A   B   C   D
nondirectional.   B) must be directional.   C) may be nondirectional or
directional.   D) is not stated.

In the Kruskal-Wallis test, the alternative hypothesis ...   A) must be   8.   A   B   C   D
nondirectional.   B) must be directional.   C) may be nondirectional or
directional.   D) is not stated.

When there are ten or more pairs of scores, the sign test employs which   9.   A   B   C   D
of the following statistics?   A) χ^2 B) F C) t D) z

Which test assumes that sampling was random?   A) The F-test.   10.   A   B   C   D
B) The Mann-Whitney test.   C) The sign test.   D) All three do.

DATA FOR USE IN HOMEWORK EXERCISES

In many locations across America, the National Oceanic and Atmospheric Administration of the U. S. Commerce Department categorizes each day as clear, partly cloudy, or cloudy. These terms refer to the sky cover during daylight hours. "Clear" means zero to 0.3 sky cover; "partly cloudy" means 0.4 to 0.7 sky cover; "cloudy" means 0.8 to complete sky cover. *The World Almanac and Book of Facts* lists the annual count of each kind of day for many cities. Listed below is the count of the number of clear days in 1979 (the latest year for which data are available as of this writing) for 40 cities.

95	Atlanta	82	Minneapolis-St. Paul
104	Baltimore	100	Nashville-Davidson
103	Boston	95	Newark*
58	Buffalo	103	New Orleans
70	Chicago	100	New York*
83	Cincinnati	87	Philadelphia
59	Cleveland	201	Phoenix
64	Columbus	62	Pittsburg
133	Dallas-Fort Worth	73	Portland OR
132	Denver-Boulder	95	Providence
80	Detroit	55	Rochester
76	Honolulu	190	Sacramento*
111	Houston	96	St. Louis
84	Indianapolis	135	Salt Lake City
114	Kansas City KS and MO	99	San Antonio
143	Los Angeles-Long Beach	140	San Diego
87	Louisville	153	San Francisco-Oakland
123	Memphis	63	Seattle
72	Miami	104	Tampa-St. Petersburg
80	Milwaukee	88	Washington DC

*For Newark, New York, and Sacramento, the 1979 count was not available. For these cities, the mean count over recent prior years is given.

In the fall of 1977, Ramapo College of New Jersey offered a statistics course (taught by someone other than the author of your *Guide* and using a text other than Minium and Clarke's) in which the students were tested with a total of 500 multiple-choice items over the semester. Eleven students completed all the work for the course, and here for ten of them are their scores on the first examination, expressed as a percentage correct out of the 39 items on that exam, along with their percentage correct out of the semester's total of 500 items. The student at the median on the first test was dropped from the table to reduce the *n* to 10 and thus simplify the calculations the homework asks you to do. In the homework for Ch. 8, you will have an opportunity to predict his score over the semester.

Student	% Correct on Exam 1	% Correct Over Semester
A	38	75
B	54	65
C	62	94
D	67	81
E	67	84
F	72	93
G	77	90
H	77	93
I	82	90
J	85	95

ANSWERS FOR CH. 1

Review

1. inferential 2. descriptive 3. descriptive 4. inferential 5. chance 6. sampling 7. relationship 8. descriptive 9. inferential 10. organize 11. summarize 12. inference 13. descriptive 14. inferential 15. part 16. population 17. chance 18. description 19. chance 20. random 21. random 22. variation 23. sampling 24. variation 25. small 26. smaller 27. variable 28. qualitative 29. qualitative 30. quantitative 31. quantitative 32. numerically 33. scaled 34. qualitative 35. substantive 36. descriptive 37. statistical 38. conclusion 39. substantive 40. exact 41. no 42. approximate 43. approximate

Self-Test

Answers to all self-tests appear in a table following the answers for the last chapter of this *Guide*.

ANSWERS FOR CH. 2

Review

1. distribution 2. table 3. lowest 4. descending 5. *f* 6. grouped 7. grouped frequency 8. grouped 9. grouped 10. grouped frequency 11. interval 12. class 13. limits 14. should not 15. the same 16. continuous 17. top 18. 20 19. convenient 20. multiple 21. relative 22. percentages 23. number 24. *f/n* 25. *f* 26. *n* 27. right 28. relative 29. limits 30. exact 31. limits 32. measurement 33. above 34. below 35. cumulative 36. upper 37. *cum f* 38. *n* 39. *cum f* 40. *n* 41. right 42. quantitative 43. numerically 44. number 45. top

Exercises

Score Limits	Exact Limits	*f*	*cum f*	*cum % f*
18–19	17.5–19.5	3	12	100
16–17	15.5–17.5	3	9	75
14–15	13.5–15.5	2	6	50
12–13	11.5–13.5	1	4	33
10–11	9.5–11.5	3	3	25

Score Limits	Exact Limits	f	cum f	cum % f
30–34	29.5–34.5	4	12	100
25–29	24.5–29.5	2	8	67
20–24	19.5–24.5	2	6	50
15–19	14.5–19.5	0	4	33
10–14	9.5–14.5	4	4	33

Score Limits	Exact Limits	f	cum f	cum % f
120–129	119.5–129.5	5	15	100
110–119	109.5–119.5	3	10	67
100–109	99.5–109.5	2	7	47
90–99	89.5 – 99.5	5	5	33

Self-Test

Answers to all self-tests appear in a table following the answers for the final chapter in this *Guide*.

ANSWERS FOR CH. 3

Review

1. grouped   2. 100   3. histogram   4. exact   5. percentage   6. horizontal   7. X 8. vertical   9. Y 10. zero   11. do not   12. three-quarters   13. labels   14. polygon   15. midpoints   16. half   17. exact   18. frequency   19. relative   20. zero   21. frequency   22. zero   23. polygon   24. table   25. table   26. distribution   27. curve   28. percentage   29. percentages   30. 100   31. upper   32. limit   33. upper   34. ogive   35. center   36. percentile   37. below   38. percentile   39. 20   40. centile   41. centile rank   42. cumulative percentage curve (ogive)   43. polygons   44. percentage   45. relative   46. equal   47. frequency   48. frequency   49. regular   50. irregular   51. greater   52. few   53. different ways   54. frequency   55. cumulative   56. central   57. variability   58. shape   59. frequency polygon   60. bell   61. distribution   62. middle   63. skewed   64. **J**   65. bimodal   66. quantitative   67. bar diagram   68. histogram

Self-Test

Answers to all self-tests appear in a table following the answers for the last chapter in this *Guide*.

ANSWERS FOR CH. 4

Review

1. average   2. location   3. greatest   4. *Mo*   5. more than   6. tendency   7. half
8. C_{50} 9. median   10. middle   11. halfway   12. *Mdn*   13. only   14. how far
away   15. half   16. right   17. highest   18. area   19. frequency   20. central
21. arithmetic   22. sum   23. number   24. midpoint   25. X 26. $\overline{X}$ 27. $\Sigma X/n$
28. sum   29. n 30. only   31. is   32. balance point   33. mean   34. negative
35. positive   36. positive   37. zero   38. $\Sigma(X - \overline{X}) = 0$ 39. median   40. mode
41. mean   42. mean   43. most   44. variation   45. toward   46. skewed   47.
median   48. open   49. central tendency   50. mode

Catch Your Own Errors

1. No   2. Yes   3. No   4. No   5. Yes

Self-Test

Answers to all self-tests appear in a table following the answers for the last chapter of this *Guide*.

ANSWERS FOR CH. 5

Review

1. variability   2. tendency   3. variability   4. chance   5. does not specify   6.
does not provide   7. no   8. difference   9. lowest   10. four   11. one-fourth   12.
Q_1, Q_2, and Q_3 13. Q_1 14. C_{50} 15. C_{75} 16. median   17. variability   18.
third   19. 50   20. $(Q_3 - Q_1)/2$ 21. Q 22. distance   23. points   24. x 25.
$x = X - \overline{X}$ 26. mean   27. mean   28. mean   29. -5 30. below   31. 5   32.
mean   33. farther   34. larger   35. deviation   36. mean   37. squared   38. S^2
39. $\Sigma x^2/n$ 40. $\Sigma(X - \overline{X})^2/n$ 41. variance   42. is not   43. measurement   44.
inferential   45. square root   46. variance   47. description   48. S 49. $\sqrt{\Sigma x^2/n}$
50. is not   51. midpoint   52. standard deviation   53. semiinterquartile range

54. most   55. variation   56. location   57. range   58. location   59. semiinterquartile range   60. semiinterquartile range   61. is not   62. two   63. lowest   64. least   65. sampling   66. size   67. larger   68. mean   69. 68   70. standard deviations   71. 95   72. standard deviations   73. 99.7   74. majority   75. $\overline{X} \pm 3S$ 76. mean   77. mean   78. standard deviation   79. standard deviation   80. .5

Special Help with the Variance and the Standard Deviation

Raw Score, X	Deviation Score, x	Squared Deviation Score, x^2
13	$+5$	25
12	$+4$	16
9	$+1$	1
7	-1	1
6	-2	4
1	-7	49

$$\Sigma X = 48 \qquad \Sigma x = 0 \qquad \Sigma x^2 = 96$$
$$n = 6 \qquad\qquad\qquad\qquad S^2 = 96/6$$
$$\overline{X} = \Sigma X/n \qquad\qquad\qquad = 16.0$$
$$= 48/6 = 8.0 \qquad\qquad S = \sqrt{16.0} = 4.0$$

The standard deviations of the two distributions work out to be the same because this measure of variability depends on the set of deviation scores, and the two distributions have the same set of deviation scores. The set of deviation scores is the same for the two distributions because where the raw scores lie in relation to their mean is the same in the two distributions.

Raw Score, X	Deviation Score, x	Squared Deviation Score, x^2
14	$+4$	16
13	$+3$	9
10	0	0
10	0	0
8	-2	4
5	-5	25

$$\Sigma X = 60 \qquad \Sigma x = 0 \qquad \Sigma x^2 = 54$$
$$n = 6 \qquad\qquad\qquad\qquad S^2 = 54/6$$
$$\overline{X} = \Sigma X/n \qquad\qquad\qquad = 9.0$$
$$= 60/6 = 10.0 \qquad\qquad S = \sqrt{9.0} = 3.0$$

Raw Score, X	Deviation Score, x	Squared Deviation Score, x^2
14	$+4$	16
9	-1	1
9	-1	1
9	-1	1
9	-1	1

$$\Sigma X = 50 \qquad \Sigma x = 0 \qquad \Sigma x^2 = 20$$
$$n = 5 \qquad\qquad\qquad\qquad S^2 = 20/5$$
$$\overline{X} = \Sigma X/n \qquad\qquad\qquad = 4.0$$
$$= 50/5 = 10.0 \qquad\qquad S = \sqrt{4.0} = 2.0$$

Raw Score, X	Deviation Score, x	Squared Deviation Score, x^2
7	$+3$	9
7	$+3$	9
7	$+3$	9
5	$+1$	1
4	0	0
4	0	0
3	-1	1
3	-1	1
3	-1	1
2	-2	4
2	-2	4
1	-3	9

$$\Sigma X = 48 \qquad \Sigma x = 0 \qquad \Sigma x^2 = 48$$
$$n = 12 \qquad\qquad\qquad\qquad S^2 = 48/12$$
$$\overline{X} = \Sigma X/n \qquad\qquad\qquad = 4.0$$
$$= 48/12 = 4.0 \qquad\qquad S = \sqrt{4.0} = 2.0$$

Squared RAW Score, X^2	Raw Score, X
169	13
144	12
81	9
49	7
36	6
1	1
$\Sigma X^2 = 480$	$\Sigma X = 48 \quad n = 6$

$\Sigma x^2 = \Sigma X^2 - (\Sigma X)^2/n$
$= 480 - (48)^2/6$
$= 480 - 2304/6$
$= 480 - 384$
$= 96$

Squared RAW Score, X^2	Raw Score, X
196	14
169	13
100	10
100	10
64	8
25	5
$\Sigma X^2 = 654$	$\Sigma X = 60 \quad n = 6$

$\Sigma x^2 = \Sigma X^2 - (\Sigma X)^2/n$
$= 654 - (60)^2/6$
$= 654 - 3600/6$
$= 654 - 600$
$= 54$

Squared RAW Score, X^2	Raw Score, X
196	14
81	9
81	9
81	9
81	9
$\Sigma X^2 = 520$	$\Sigma X = 50 \quad n = 5$

$\Sigma x^2 = \Sigma X^2 - (\Sigma X)^2/n$
$= 520 - (50)^2/5$
$= 520 - 2500/5$
$= 520 - 500$
$= 20$

Squared RAW Score, X^2	Raw Score, X	
49	7	$\Sigma x^2 = \Sigma X^2 - (\Sigma X)^2/n$
49	7	$= 240 - (48)^2/12$
49	7	$= 240 - 2304/12$
25	5	$= 240 - 192$
16	4	$= 48$
16	4	
9	3	
9	3	
9	3	
4	2	
4	2	
1	1	
$\Sigma X^2 = 240$	$\Sigma X = 48 \quad n = 12$	

Descriptive Statistics in Use

1. The class interval containing the largest scores is on the bottom. There are too few intervals. The intervals are not of uniform width. One interval is open-ended, but the distributions are not open-ended, for all numbers over 25 are known.   2. Skewed to the right.   3. Skewed to the right.   4. 3.2   5. 2.0   6. The very large scores pull the mean up. See the last paragraph on p. 59 of the text.   7. For the same reason as in the first distribution.   8. $(7.6-0)/2 = 3.8$ 9. $(4.2-0)/2 = 2.1$ 10. Because the distributions are so highly skewed, the standard deviation S would be misleadingly large. See the second paragraph on p. 75 of the text.   11. Since $\bar{X} = \Sigma X/n$, $\Sigma X = n\bar{X}$. Here $n\bar{X} = (176)(6.81) = 1198.56$. The total must have been a whole number and could have been either 1198 or 1199; both figures yield a mean that rounds to 6.81 (when the figure is divided by 176).   12. Again we must compute $n\bar{X}$. Here the figures are $(128)(2.97) = 380.16$. Since the total must have been a whole number, it would have been 380.   13. The researchers sought to study the entire set of American professionals who did diagnostic cytogenetics and genetic counseling. The populations of interest to the researchers were the sets of answers to the items on the questionnaire from this set of people. Since all but five of the people responded to the questionnaire, each sample of answers was almost a complete

population. When so few observations are missing from a sample, its characteristics are almost certainly very close to those of its parent population and there is no need for inferential techniques to draw a conclusion about the population.

ANSWERS FOR CH. 6

Review

1. shape   2. curve   3. statistics   4. inferential   5. normal   6. frequency polygon   7. 68   8. 95   9. mean   10. 99.7   11. area   12. polygon   13. 68   14. 95   15. $\overline{X} \pm 3S$ 16. symmetrical   17. unimodal   18. mode   19. horizontal   20. X 21. 99.7   22. $\overline{X} \pm 3S$ 23. 700   24. 800   25. 70   26. 55   27. z 28. z 29. mean   30. standard deviation   31. deviation   32. mean   33. -1 34. $+2$ 35. 145   36. 200   37. mean   38. standard deviation   39. $(X - \overline{X})/S$ 40. z 41. shape   42. zero   43. one   44. z 45. z 46. mean   47. mean   48. deviation   49. reference   50. reference   51. negative   52. above   53. mean   54. below   55. mean   56. 500   57. 100   58. reference   59. 63   60. reference group   61. 50   62. 84   63. 98   64. 99.9   65. centile ranks   66. the center   67. an extreme

Exercises

TABLE OF EQUIVALENT SCORES

z Score	Score where $\overline{X}=100, S=15$	Score where $\overline{X}=500, S=100$	Score where $\overline{X}=50, S=10$	Centile Rank if Shape is Normal
$+3.00$	145	800	80	99.87
$+2.00$	130	700	70	97.72
$+1.50$	122.5	650	65	93.32
0.00	100	500	50	50.00
-1.00	85	400	40	15.87
-2.50	62.5	250	25	0.62
-3.00	55	200	20	0.13

Special Properties of a Distribution of z scores

Raw Score	Deviation Score	Squared Deviation	z Score	Deviation Score for z Score	Squared Deviation for z Score
X	$x=(X-\bar{X})$	x^2	$z=x/S$	$(z-\bar{z})$	$(z-\bar{z})^2$
13	$+5$	25	$+1.25$	$+1.25$	1.5625
13	$+5$	25	$+1.25$	$+1.25$	1.5625
9	$+1$	1	$+0.25$	$+0.25$	0.0625
6	-2	4	-0.50	-0.50	0.2500
4	-4	16	-1.00	-1.00	1.0000
3	-5	25	-1.25	-1.25	1.5625

$\Sigma X=48$ $\qquad$ $\Sigma x=0$ $\qquad$ $\Sigma x^2=96$ $\qquad$ $\Sigma z=0.00$ $\qquad$ $\Sigma(z-\bar{z})=0.00$ $\qquad$ $\Sigma(z-\bar{z})^2=6.0000$

$n=6$ $\qquad\qquad\qquad\qquad\qquad\qquad$ $S^2=\Sigma x^2/n$ $\qquad$ $n=6$ $\qquad\qquad\qquad\qquad\qquad$ $S_z^2=\Sigma(z-\bar{z})^2/n$

$\bar{X}=\Sigma X/n$ $\qquad\qquad\qquad\qquad\qquad\quad$ $=96/6$ $\qquad$ $\bar{z}=\Sigma z/n$ $\qquad\qquad\qquad\qquad$ $=6.0000/6$

$=48/6$ $\qquad\qquad\qquad\qquad\qquad\qquad$ $=16.0$ $\qquad$ $=0.00/6$ $\qquad\qquad\qquad\qquad$ $=1.0000$

$=8.0$ $\qquad\qquad\qquad\qquad\qquad\qquad\;$ $S=\sqrt{16.0}$ $\qquad$ $=0.00$ $\qquad\qquad\qquad\qquad\quad$ $S_z=\sqrt{1.0000}$

$\qquad\qquad\qquad\qquad\qquad\qquad\qquad\qquad$ $=4.0$ $\qquad\qquad\qquad\qquad\qquad\qquad\qquad\qquad$ $=1.0000$

The Normal Distribution and z Scores in Use

1. $100-15=85$. See p. 105 of the text for a picture of the distribution.   2. About 16%.   3. $70-100=-30$, which is 15×-2. Therefore 70 is two standard deviations below the mean.   4. About 2%.   5. $50-100=-50$, which is 15×-3.33. Therefore 50 is 3.33 standard deviations below the mean.   6. According to Appendix F, 0.05% of the cases lie beyond a z score of 3.30, plus or minus. (The value of 3.30 is the closest we can get to 3.33 in the table.) Coming in toward the mean to a z of -2.00 (corresponding to an IQ of 70), we find that 2.28% of the cases lie beyond it. That leaves $2.28-0.05=2.23\%$ of the cases in the interval between $z=-3.30$ and $z=-2.00$. So about 2% of the population is mildly retarded by Zigler's definition.   7. Somewhat less than 0.05 (which is the figure for a z score of -3.30). This is about 1 person in 2000.

ANSWERS FOR CH. 7

Review

1. paired   2. X 3. Y 4. bivariate   5. distribution   6. scatter diagram   7. bivariate   8. Y 9. diagrams   10. distributions   11. negative   12. high   13. low   14. high   15. low   16. upper right   17. low   18. high   19. low   20. high   21. lower right   22. diagram   23. straight   24. upper right   25. lower   26. coefficient   27. r 28. distribution   29. negative   30. high   31. low   32. negative   33. low   34. direction   35. 0   36. 1   37. r 38. straight   39. $\Sigma xy/nS_X S_Y$ 40. $(Y - \overline{Y})$ 41. deviation   42. pairs   43. standard deviation   44. S_Y 45. scatter   46. mean   47. mean   48. positive   49. negative   50. above   51. below   52. I and III   53. II and IV   54. sign   55. I and III   56. II and IV   57. II and IV   58. I and III   59. Σxy 60. negative   61. r 62. causal   63. Y 64. X 65. correlation   66. regression   67. curvilinear   68. underestimate   69. coefficient   70. lower   71. r 72. relative   73. Y 74. chance   75. less   76. restricted   77. chance   78. subjects   79. zero   80. one   81. zero   82. one   83. straight   84. less   85. scatter diagram   86. variance   87. deviation   88. square

Estimating Correlations

1. Positive and high, surely very close to perfect (1.00).   2. Positive and probably at least medium. The older children will have longer noses and larger vocabularies, so a child with a high score on one variable will generally have a high score on the other, and a child with a low score on one variable (who is likely to be a younger child) will generally have a low score on the other variable. The examples in the first two questions here show that two variables can be correlated even though neither has an influence on the other.   3. Almost certainly zero.   4. Positive and probably high. Note that instead of two scores for a single subject, here we have two scores for a pair of subjects (a couple). The couple is thus the equivalent of a single subject, in that it is the unit on which the two variables are measured.   5. Still positive and high. The change in social custom would not influence the *relationship* between the two variables; it would only raise the scores on one variable (husband's age) relative to the scores on the other variable (wife's age). A couple with a score for husband's age that is high for scores of this kind will still tend to have a score for wife's age that is high for scores of *its* kind. Similarly, a couple with a score for husband's age that is low relative to other such scores will tend to have a score for wife's age that is low relative to the other scores for the wives. 6. The question is nonsensical, because there are no pairs of scores here. There is no logical way to pair a baseball player's height with a football player's height; the two teams probably have different

numbers of players, for one thing. So the concept of correlation does not apply to a situation like this. (Life is full of questions like this one, questions to which the proper answer is "That's a stupid question." Stay alert for them.)

ANSWERS FOR CH. 8

Review

1. scatter diagram 2. Y 3. X 4. vertical 5. linearity 6. X 7. horizontal 8. Y 9. vertical 10. Y' 11. regression 12. -1 13. exactly on 14. $Y - Y'$ 15. sum 16. squares 17. mean 18. regression equation 19. predicted 20. standard deviation 21. S_X 22. mean 23. $\overline{Y}$ 24. Y' 25. mean 26. rz_X 27. correlation 28. regression 29. identical 30. -2 31. standard deviations 32. one 33. mean 34. 0 35. mean 36. mean 37. zero 38. regression 39. error 40. S_{YX} 41. larger 42. variability 43. $\sqrt{\Sigma(Y - \overline{Y})^2/n}$ 44. Y' 45. $\overline{Y}$ 46. prediction 47. $S_Y\sqrt{1 - r^2}$ 48. mean 49. 1.96 50. mean 51. $(1.96)S_{YX}$ 52. population 53. small 54. 100 55. larger 56. smaller 57. S_{YX} 58. 0 59. mean 60. $\overline{Y}$ 61. S_{YX} 62. S_Y 63. 3 64. 34 65. strong

Patterns in Predictions

	If r is positive	If r is zero	If r is negative
If $X > \overline{X}$	$Y' > \overline{Y}$	$Y' = \overline{Y}$	$Y' < \overline{Y}$
If $X = \overline{X}$	$Y' = \overline{Y}$	$Y' = \overline{Y}$	$Y' = \overline{Y}$
If $X < \overline{X}$	$Y' < \overline{Y}$	$Y' = \overline{Y}$	$Y' > \overline{Y}$

ANSWERS FOR CH. 9

Review

1. chance 2. sample 3. chance 4. probability 5. experiment 6. proportion 7. experiments 8. proportion 9. cards 10. proportion 11. proportion 12. table 13. percentage 14. probabilities 15. proportion 16. probability 17. same 18. above the mean 19. zero 20. scores 21. proportion 22. area 23. exclusive 24. cannot 25. mutually 26. cannot 27. mutually exclusive 28. sum 29. plus 30. exclusive 31. add 32. independent 33. independent 34. product 35. times 36. multiply

ANSWERS FOR CH. 10

Review

1. sampling   2. population   3. population   4. sample   5. randomization   6. different   7. sampling   8. randomization   9. model   10. complete   11. part   12. parameters   13. statistics   14. μ 15. σ 16. $\bar{X}$ 17. S 18. the same   19. sample   20. No   21. random   22. random   23. random   24. $\bar{X}$ 25. random   26. distribution   27. sampling   28. frequency   29. random   30. deviation   31. shape   32. means   33. $\mu_{\bar{X}}$ 34. population   35. $\mu_{\bar{X}}$ 36. μ 37. error   38. $\sigma_{\bar{X}}$ 39. mean   40. deviation   41. $\sigma/\sqrt{n}$ 42. less   43. more   44. normally   45. increases   46. Limit   47. sampling   48. distribution   49. z 50. mean   51. standard deviation   52. score   53. mean   54. error   55. $(\bar{X} - \mu)/\sigma_{\bar{X}}$ 56. all   57. sampling   58. relative   59. population

ANSWERS FOR CH. 11

Review

1. error   2. $\sigma_{\bar{X}}$ 3. $\sigma/\sqrt{n}$ 4. standard   deviation   5. S 6. small   7. $\sqrt{\Sigma x^2/(n-1)}$ 8. $\sqrt{\Sigma x^2/n}$ 9. $n - 1$ 10. larger   11. standard   12. mean   13. $s_{\bar{X}}$ 14. $s/\sqrt{n}$ 15. null   16. H_0 17. reject   18. H_A 19. null   20. hypothesis   21. null   22. alternative   23. mean   24. greater   25. null   26. alternative   27. before   28. two-tailed   29. either   30. alternative   31. one-tailed   32. H_0 33. reject   34. .01   35. H_0 36. significance   37. null   38. $\mu_{\bar{X}}$ 39. μ_{hyp} 40. null   41. 126   42. means   43. rejection   44. null   45. reject   46. acceptance   47. rejection   48. lower   49. right   50. significance   51. critical   52. α 53. H_A 54. rejection   55. mean   56. mean   57. error   58. $s_{\bar{X}}$ 59. mean   60. null   61. null   62. reject   63. reject   64. alternative   65. null   66. sampling   67. significance   68. large

ANSWERS FOR CH. 12

Review

1. population   2. mean   3. accept   4. reject   5. true   6. means   7. $\bar{X}$ 8. sample   9. null   10. error   11. $\sigma_{\bar{X}}$ 12. $s_{\bar{X}}$ 13. t 14. $\sigma_{\bar{X}}$ 15. z 16. z 17. $\bar{X}$ 18. s 19. standard error   20. mean   21. deviation   22. 0   23. more   24. 0   25. less   26. more   27. normal   distribution   28. freedom   29. degrees   30. in-

ference   31. σ 32. standard   error   33. $n-1$ 34. df 35. t 36. normal   37. normal   38. t 39. α 40. significance   41. .10   42. .01   43. H_0 44. true   45. α 46. rejects   47. accepts   48. $>$ 49. I   50. α 51. .05   52. significance   53. rejecting   54. reject   55. II   56. I   57. .05   58. .01   59. I   60. II

ANSWERS FOR CH. 13

Review

1. population   2. chance   3. population means   4. population   5. μ_2 6. means
7. less than   8. two   9. two   10. $-\mu_2 < 0$ 11. population   12. $(\overline{X}_1 - \overline{X}_2)$
13. true   14. accept   15. reject   16. means   17. α 18. greater   19. less   20. $(\overline{X}_1 - \overline{X}_2)$ 21. true   22. two   means   23. sampling   24. $(\overline{X}_1 - \overline{X}_2)$ 25. $(\overline{X}_1 - \overline{X}_2)$ 26. sampling   27. random   sampling   28. differences   29. $\mu_{\overline{X}_1 - \overline{X}_2}$
30. error   31. independent   32. random   33. standard   34. $\sigma_{\overline{X}_1}$ 35. $\sigma_{\overline{X}_2}$ 36. $\sigma_{\overline{X}_1 - \overline{X}_2}$ 37. $\sqrt{\sigma_{\overline{X}_1}^2 + \sigma_{\overline{X}_2}^2}$ 38. smaller   39. larger   40. normal   41. Central Limit
42. n_1 and n_2 43. $\sigma_{\overline{X}_2}$ 44. $\sigma_{\overline{X}_2}^2$ 45. difference   46. σ^2 47. standard deviation
48. freedom   49. variance   50. population   51. s_p^2 52. $(\Sigma x_1^2 + \Sigma x_2^2)/(n_1 + n_2 - 2)$ 53. $s_{\overline{X}_1 - \overline{X}_2}$ 54. $\sqrt{s_p^2/n_1 + s_p^2/n_2}$ 55. true   56. $(\overline{X}_1 - \overline{X}_2)$ 57. t 58. $[(\overline{X}_1 - \overline{X}_2) - 0]/s_{\overline{X}_1 - \overline{X}_2}$ 59. df 60. $n_1 + n_2 - 2$ 61. significance   62. alternative
63. accept   64. null   65. level   66. large   67. standard deviation   68. sample
69. $(\overline{X}_1 - \overline{X}_2)$ 70. population   71. variance   72. shape   73. variability   74. randomization   75. chance   76. extraneous   77. unlikely   78. does not guarantee   79. increases   80. randomly   81. population   82. substantive   83. inferential   84. randomization   85. nonstatistical

ANSWERS FOR CH. 14

Review

1. related   2. both   3. two scores   4. can also   5. independent   6. error   7. s_p^2
8. $\sqrt{s_p^2/n_1 + s_p^2/n_2}$ 9. correlation   10. r 11. high   12. low   13. subtracted from   14. smaller   15. $\sqrt{s_1^2/n_1 + s_2^2/n_2 - 2r_{12}s_1s_2/n}$ 16. estimate   17. variance
18. standard deviation   19. s^2 20. pairs of scores   21. $s_{\overline{X}_1 - \overline{X}_2}$ 22. r 23. $s_{\overline{X}_1 - \overline{X}_2}$
24. df 25. $n-1$ 26. dependent   27. differences   28. $\mu_1 - \mu_2 = 0$ 29. zero
30. alternative   31. $<$ 32. D 33. X 34. $s_{\overline{D}}$ 35. sample   36. null   37. D
38. $s_{\overline{X}_1 - \overline{X}_2}$ 39. the same   40. $n-1$ 41. less   42. r 43. standard error   44. means   45. randomization   46. randomization   47. matched   48. order

Annals of Egregious Examples

1. Calling the two samples X_1 and X_2, you should compute $\overline{X}_1$, s_1, $\overline{X}_2$, s_2, and $\overline{X}_1 - \overline{X}_2$. You should compare the latter to the mean of s_1 and s_2, as Section 5.8 on pp. 78–80 advises, so you get some idea of how large the difference between the sample means is.   2. You should test a hypothesis about the difference between two population means. The samples are independent. The null should state that the difference between the two population means is zero, and the alternative should say that it is not zero, which is the two-tailed case. You would have to choose an α level, estimate $\sigma_{\overline{X}_1 - \overline{X}_2}$, and calculate the t ratio.   3. You should proceed as for the first study (see Question 1), and in addition you should calculate the correlation coefficient r for the two samples of scores.   4. This is a case of dependent means, so you have your choice of the procedures described in Sections 14.4 and 14.5. The null hypothesis should again declare no difference between the two population means, or a mean difference score of zero in the population of such scores. The alternative should again be two-tailed. You would have to calculate $s_{\overline{X}_1 - \overline{X}_2}$ or $s_{\overline{D}}$ and the t ratio again.   5. In the second study, each score for Variety A was paired with a score for Variety B, but the tables of data do not indicate the pairings. It is thus impossible to compute $s_{\overline{X}_1 - \overline{X}_2}$ or the difference scores and $s_{\overline{D}}$. Hence there is no way to determine the value of the t ratio and the plausibility of the null hypothesis.

ANSWERS FOR CH. 15

Review

1. false 2. rejection 3. such 4. $\mu_1 - \mu_2 = 0$ 5. variance 6. σ_2^2 7. variance 8. σ 9. $(\mu_1 - \mu_2)/\sigma$ 10. effect size 11. error 12. d 13. greater 14. $\mu_1 - \mu_2 = 0$ 15. $\mu_1 - \mu_2 \neq 0$ 16. in 17. $\sigma_{\overline{X}_1 - \overline{X}_2}$ 18. smaller 19. $s_{\overline{X}_1 - \overline{X}_2}$ 20. smaller 21. larger 22. 0 23. more 24. less 25. large 26. α 27. rejection 28. I 29. independent 30. difference 31. correlation 32. larger 33. increase 34. correlation 35. alternative 36. increase 37. two-tailed 38. .05 39. higher 40. correlation 41. lower 42. rejected 43. significant 44. nonsignificant 45. d 46. probability 47. reject 48. accept 49. level of significance 50. higher 51. higher 52. lower 53. rejecting 54. accepting 55. $1 -$ power

ANSWERS FOR CH. 16

Review

1. null 2. μ 3. parameters 4. parameter 5. estimation 6. interval 7. point 8. statistics 9. chance 10. interval 11. wide 12. less 13. 95 14. 1.96 15.

95   16. μ 17. standard error   18. $\sigma_{\bar{X}_1 - \bar{X}_2}$ 19. σ 20. $\sigma_{\bar{X}}$ 21. $s/\sqrt{n}$ 22. t 23.
95   24. both tails combined   25. $n-1$ 26. t 27. 1   28. the same as   29. $(\bar{X}_1$
$- \bar{X}_2)$ 30. $n_1 + n_2 - 2$ 31. $n-1$ 32. wider   33. size   34. more   narrow   35.
.05   36. .01   37. accepted   38. estimation   39. normally   40. variance   41.
small

ANSWERS FOR CH. 17

Review

1. sample   2. sampling   3. parameter   4. estimation   5. zero   6. ρ 7. $\rho = 0$
8. $\rho \neq 0$ 9. <0 10. significance   11. df 12. $n-2$ 13. null   14. critical   15.
alternative   16. n 17. $n-2$ 18. 0   19. ρ 20. power   21. population   22. ρ
23. powerful   24. ρ 25. significant   26. interval   27. limits

ANSWERS FOR CH. 18

Review

1. power   2. significant   3. one   4. independent   5. t 6. independent   7. at
random   8. $D, E, F, \ldots$ 9. k 10. $\mu_D, \mu_E, \mu_F, \ldots$ 11. mean   12. $\mu_D = \mu_E = \mu_F$
$= \ldots$ 13. no   longer   makes   14. $\bar{X}_D$ 15. $\bar{X}_E$ 16. $\bar{X}_F$ 17. within-groups   18.
inherent   19. among-groups   20. greater   21. within   22. is   not   23. inherent
24. treatment   25. among   26. within   27. inherent   28. within-groups   29.
among-groups   30. accept   31. inherent   32. within-groups   33. reject   34. ho-
mogeneity   35. σ_D^2 36. σ_E^2 37. σ_F^2 38. σ^2 39. s^2 40. sum of squares   41. SS
42. SS/df 43. s_W^2 44. SS_W 45. within   46. df_W 47. SS_W 48. $X - \bar{X}$ 49.
$\Sigma(X - \bar{X})^2$ 50. $\bar{X}_D$ 51. $\bar{X}_E$ 52. sample   size   53. k 54. $n_T - k$ 55. s_W^2 56.
SS_W/df_W 57. σ^2 58. s_A^2 59. squares   60. SS_A 61. df_A 62. $\bar{X}$ 63. grand
64. $\Sigma(\bar{X} - \bar{\bar{X}})^2$ 65. groups (samples)   66. $k-1$ 67. SS_A/df_A 68. among   69.
F 70. s_A^2/s_W^2 71. true   72. equal   73. t 74. t 75. df_A 76. df_W 77. false
78. inherent   79. t 80. independent   81. small   82. SS_T 83. deviations   84.
grand   85. $\Sigma(X - \bar{X})^2$ 86. df_T 87. $n_T - 1$ 88. degrees of freedom   89. reject
90. planned   91. post hoc   92. honest significant difference   93. equal   94. reject

Annals of Egregious Examples, *continued*

1. A one-way analysis of variance, the kind of analysis described in this chapter.   2.
No, because the samples are dependent.

ANSWERS FOR CH. 19

Review

1. 60 2. 20 3. 20 4. 20 5. column 6. factor 7. levels 8. factor 9. column 10. one-way 11. column 12. 10 13. 10 14. two 15. row 16. two-factor 17. row 18. column 19. cell 20. 6 21. 10 22. cell 23. independent 24. cell 25. 30 26. row 27. $\overline{X}_H$ 28. equal 29. 20 30. column 31. $\overline{X}_A$ 32. cell means 33. grand 34. analysis of variance 35. $\mu_A = \mu_B = \mu_C$ 36. $\overline{X}_A$, $\overline{X}_B$, and $\overline{X}_C$ 37. main effect 38. $\mu_A = \mu_B = \mu_C$ 39. row factor 40. column factor 41. levels 42. $\mu_H = \mu_L$ 43. false 44. interaction 45. column 46. the same 47. yes 48. no 49. an 50. worse 51. 84 52. 92 53. no 54. populations 55. column factor 56. interaction 57. F 58. larger 59. inherent 60. one-way 61. within 62. s_W^2 63. within-cells 64. s_{WC}^2 65. among 66. s_A^2 67. three 68. main 69. s_R^2 70. s_C^2 71. $s_{R \times C}^2$ 72. squares 73. degrees of freedom 74. four 75. grand 76. $\Sigma(X - \overline{\overline{X}})^2$ 77. SS_T 78. cell 79. $\Sigma(X - \overline{X}_{\text{cell}})^2$ 80. SS_{WC} 81. grand 82. cell 83. $SS_{R \times C}$ 84. one 85. equal 86. R 87. cells 88. $(R)(C)(n_{\text{cell}}) - 1$ 89. $k - 1$ 90. $R - 1$ 91. $C - 1$ 92. product 93. $(R - 1)(C - 1)$ 94. estimate 95. critical 96. degrees of freedom 97. within-cells 98. reject 99. homogeneity 100. normally 101. large 102. null 103. Tukey HSD 104. three 105. two 106. interaction

ANSWERS FOR CH. 20

Review

1. dichotomous 2. P 3. two 4. $1 - P$ 5. n 6. 0 7. n 8. 1/4 9. 1/4 10. 4 11. 0 12. 4 13. 4 14. .3164 15. .4219 16. .0039 17. .3164 18. one 19. 1/4 20. population 21. population 22. 4 23. 0 24. 4 25. histogram 26. sampling 27. count 28. 4 29. 1/4 30. 0, 1, 2, 3, and 4 31. proportion 32. μ 33. binomial 34. .50 35. n 36. 15 37. X 38. n 39. P 40. .50 41. n 42. 15 43. 1/4 44. 7 45. mutually 46. the same as 47. null 48. P 49. H_A 50. $\neq$ 51. two 52. $<$ 53. α 54. n 55. statistic 56. X 57. sampling 58. null hypothesis 59. alternative 60. greater 61. smaller 62. α 63. null 64. count 65. skewed 66. two 67. rejection 68. P 69. remains the same

ANSWERS FOR CH. 21

Review

1. numerically 2. qualitative 3. frequency 4. category 5. four 6. proportions 7. population 8. null 9. H_0 10. need not 11. f_e 12. *true* 13. null 14. n 15. observed 16. f_o 17. $\Sigma f_o = n$ 18. cannot 19. category 20. more 21. 1/4 22. true 23. false 24. accept 25. reject 26. χ^2 27. expected 28. observed 29. $f_o - f_e$ 30. $(f_o - f_e)^2$ 31. expected 32. $(f_o - f_e)^2/f_e$ 33. $\Sigma[(f_o - f_e)^2/f_e]$ 34. expected frequency 35. larger 36. does not matter 37. true 38. number 39. expected 40. freedom 41. $C - 1$ 42. sample 43. the same as 44. true 45. accept 46. reject 47. significance 48. .05 49. sampling 50. true 51. degrees of freedom 52. skewed right 53. proportion 54. population 55. false 56. .05 or .01 57. at random 58. freedom 59. upper 60. larger 61. less than 62. reject 63. dichotomous 64. χ^2 65. proportion 66. 1 67. directional 68. alternative 69. row 70. columns 71. boxes 72. observed 73. contingency 74. contingency 75. f_{row} 76. $p_o = f_o/f_{row}$ 77. independent 78. dependent 79. .45 80. .25 81. nonvoting 82. independent 83. is the same for all 84. two 85. f_{col} 86. n 87. total 88. 1.00 89. observed frequencies 90. times 91. columns 92. $(R - 1)(C - 1)$ 93. sampling 94. small 95. independent 96. 100 97. top 98. 1 99. one 100. random 101. independent 102. wrong 103. large 104. small

ANSWERS FOR CH. 22

Review

1. t 2. F 3. t 4. normal 5. variance 6. large 7. population 8. nonparametric 9. "significant" 10. powerful 11. mean 12. central tendency 13. reject 14. variability and shape 15. rank 16. 1 and 2 17. mean 18. 1.5 19. one-third 20. scores 21. even if there are some 22. Mann-Whitney test 23. all ways 24. central tendency 25. median 26. nondirectional 27. X 28. lowest 29. lowest 30. X 31. ΣR_X 32. directional or nondirectional 33. z 34. normal 35. random 36. will not 37. Kruskal-Wallis 38. t 39. Kruskal-Wallis 40. Mann-Whitney 41. locations 42. median 43. many ways 44. lowest 45. ΣR_1 46. ΣR_2 47. χ^2 48. groups 49. upper 50. four 51. Mann-Whitney 52. random 53. ties in rank 54. sign 55. n 56. population 57. unequal 58. minuses 59. null 60. χ^2 61. observed 62. half 63. proportion 64. f_e 65. χ^2 66. 1 67. nondirectional 68. directional 69. binomial 70. random 71. n

ANSWERS TO SELF-TESTS

CHAPTER

QUESTION	1	2	3	4	5	6	7	8	9	10	11	12	13	14	15	16	17	18	19	20	21	22
1	B	D	D	D	D	C	B	D	D	B	D	D	A	C	D	B	C	C	C	A	C	D
2	B	C	D	A	B	D	B	C	C	A	B	C	C	D	A	C	C	D	D	D	D	A
3	C	B	B	C	C	A	A	A	A	A	B	B	B	B	A	A	B	D	B	C	D	D
4	D	D	A	C	C	D	C	B	B	C	A	B	D	C	D	D	A	A	B	A	C	D
5	D	A	C	C	D	D	B	B	C	B	C	B	A	D	C	D	D	D	D	A	B	D
6	B	B	B	C	A	A	D	C	B	C	A	C	A	B	A	B	C	C	A	C	D	C
7	A	A	C	C	D	D	A	B	D	A	C	D	C	A	C	D	A	B	A	C	C	C
8	C	C	C	B	A	B	C	C	A	B	C	A	D	C	D	A	A	A	B	D	D	A
9	D	C	B	A	C	B	D	D	D	C	D	A	C	C	B	B	C	A	D	D	B	A
10	C	A	C	D	A	B	A	B	C	D	D	C	A	B	D	D	D	A	C	C	B	D

HOMEWORK

On the following pages is homework, one or two double-sided sheets for each chapter of the text. The answers to the homework appear only in the instructor's manual for the text.

Most of the problems in the homework are modeled after ones that appear in the text or in this *Guide*, to encourage you to do those for practice.

The space for your name is at the bottom of the second side of the homework sheets, you will note, and you should write your name there *upside down*. The person who checks your work is thus unlikely to know who you are until she or he has finished the checking, and there will then be no question of bias in his or her judgments.

HOMEWORK FOR CHAPTER 1

In the early 1970's, social scientists at the University of Michigan's Survey Research Center wanted to know to what extent American adults were satisfied with their lives —with specific aspects such as job, neighborhood, and family, as well as with life in general. To collect relevant observations, they hired interviewers to administer a questionnaire to 2164 people from across the nation. They selected the 2164 through a process intended to make them representative of all Americans over the age of 17; an important feature of the process was random choice.

Among the items on the questionnaire was one that asked, "How satisfied are you with your life *overall?*" For various reasons (e.g., because some interviews were interrupted), this question was put to only 2134 of the 2164 respondents. Of the 2134, 463 said "Completely satisfied." These 463 constituted 21.7% of the 2134. (463/2134=.217, and the proportion .217, when expressed as a percentage, is 21.7.)

Through some statistical techniques, the researchers calculated a margin of error for the figure 21.7. The margin of error associated with 95% confidence in their finding proved to be 2.3 percentage points. The research thus permitted this conclusion:

> If all American adults answered the question "How satisfied are you with your life overall?" the percentage of answers of the kind "Completely satisfied" would lie between 21.7−2.3 and 21.7+2.3, or between 19.4% and 24.0%—and we can be 95% confident that this assertion is correct.

(The conclusion applied only at the time of the survey, of course.)

1. In that part of the research described in the second and third paragraphs, what was the subject matter of the researchers' substantive question?

2. In that part of the research described in the second and third paragraphs, what was the population of interest to the researchers?

3. In that part of the research described in the second and third paragraphs, what was the sample?

4. Write in the size of the sample here: _____

5. Was the size of the sample an exact number or an approximate number? Circle your answer and justify it: Exact   Approximate

6. If the researchers had put the question about the overall quality of life to 4268 people selected in the way the 2134 were, would the margin of error in their finding have been smaller than 2.3 percentage points or larger? Circle your answer and justify it: Larger   Smaller

7. Of the various things the researchers did in that part of their work described in the second and third paragraphs, which are techniques of descriptive statistics?

8. Of the various things the researchers did in that part of their work described in the second and third paragraphs, which are techniques of inferential statistics?

9. Is the conclusion reported in the third paragraph a substantive one or a statistical one? Circle your answer and justify it: Substantive   Statistical

10. Is any aspect of the work described in the second and the third paragraphs concerned with relationship (in the sense in which Section 1.3 of the text uses this term)? Circle your answer, and if it's yes, say which aspect: Yes   No

11. Is any aspect of the work described in the second and third paragraphs concerned with prediction (in the sense in which Section 1.3 of the text uses this term)? Circle your answer, and if it's yes, say which aspect: Yes   No

12. The researchers noted the race to which each of their respondents belonged: "black," "hispanic," "white," or "other." Which kind of variable is race? Circle your answer:

Quantitative and numerically scaled • Quantitative but not numerically scaled
Qualitative

HOMEWORK FOR CHAPTER 2

Write the exact limits for the scores listed below.

Lower Limit   Upper Limit

_____   _____   1. The weight of a 97-pound weakling measured to the nearest pound.

_____   _____   2. The weight of a 44-kilogram weakling measured to the nearest kilogram.

_____   _____   3. The weight of a 210-pound bully measured to the nearest 10 pounds.

_____   _____   4. The weight of a 95-kilogram bully measured to the nearest 5 kilograms.

_____   _____   5. A time of 9.9 seconds for a hundred-yard dash measured to the nearest tenth of a second.

_____   _____   6. A distance of 100 yards for a 9.9-second dash measured (very crudely) to the nearest 100 yards.

7. In the appendix to this *Guide* there appear 40 scores pertaining to weather in American cities. They are in the kind of jumble that typically occurs when a researcher collects observations. Bring some order to that chaos by casting the scores into a grouped frequency distribution. Use class intervals that are 10 scores wide, and let the lowest interval be 50-59. Show your results in the table on p. 262. Extend the table by filling in the column of proportionate frequencies. Give the proportions to three decimal places (e.g., as 0.123).

Score Limits	f	prop. f
50-59		

The purpose of displaying data in a table like the one you just constructed is to make their important features easy to see. On the basis of the table, *and only on the basis of the table, without going back to the original scores*, answer the following questions about the important features of those 40 clear-day counts.

8. Over what span do the numbers range? (A precise answer is impossible on the basis of the table, of course, because you grouped the data.)

9. In what span do the bulk of the numbers lie? (Again, no precise answer is possible—this time, not only because information was lost in grouping, but also because the term "bulk" is vague.)

10. Do the numbers that fall away from the bulk (either above or below) lie close to the bulk, or distant from it? Say which, and explain.

11. Now recast the 40 scores into a cumulative percentage-frequency distribution. Use the same grouping scheme, with intervals 10 scores wide beginning with 50-59. Show your results by filling in the table below. Give the percentages to one decimal place (e.g., as 12.3).

Score Limits	Exact Limits	cum f	cum % f
50-59			

If you won't get this sheet back before you do the homework for the next chapter, keep a copy of both tables.

HOMEWORK FOR CHAPTER 3

1. For the 40 counts of clear days in 1979 that you worked with in doing thehomework for Ch. 2, make a histogram. Let the scores be grouped as you grouped them for Ch. 2, into intervals 10 scores wide beginning with 50-59. Make the histogram on the front page of the next sheet. Turn the sheet on its side. On the horizontal axis, which should be the one that runs across the longer side of the page, let one unit of the graph paper represent a span of 5 days. Remember that this axis does not have to show all the possible scores from zero to the highest that actually occurred. Let the vertical axis represent raw frequency, not relative frequency.

_____2. In your histogram, there should be a rectangle representing five scores. Over what value on the X-axis should the left side of this rectangle appear?

_____3. In your histogram, over what value on the X-axis should the right side of the rectangle representing those five scores appear?

_____4. In your histogram, there should be a rectangle representing six scores. Over what value on the X-axis should the left side of the rectangle appear?

_____5. Over what value on the X-axis should the right side of that rectangle representing six scores appear?

_____6. What proportion of the total area of all the rectangles should be occupied by the rectangle representing scores from 140 to 149? Show your work below, and answer to three decimal places (e.g., as .123.)

_____7. What *percentage* of the total area of all the rectangles should be occupied by the one representing scores from 50 to 59? Show your work below, and answer to one decimal place (e.g., as 12.3).

8. On the back page of the next sheet of paper in this *Guide*, make the frequency polygon for the 40 clear-day counts as you have been grouping them. As for the histogram you made, use the paper with the longer side running horizontally, and let one unit of the graph paper represent a span of 5 days. This time, show not raw frequency but proportionate frequency on the Y-axis. Give the proportions to three decimal places (e.g., as 0.123).

_____9. In your frequency polygon, there should be a point representing a frequency of eight for the scores in a certain class interval. Over what value on the X-axis should this point appear?

_____10. In your frequency polygon, there should be a point representing a frequency count of four for the scores in a certain class interval. Over what value on the X-axis should this point appear?

11. On the back of your frequency polygon, make the cumulative percentage frequency curve for the 40 scores, again with the scores grouped into class intervals 10 scores wide beginning with 50-59. Again, use the paper with the longer axis serving as the horizontal one, and again let one unit represent a span of five days.

_____12. In your cumulative percentage frequency curve, there should be a point representing a cumulative percentage of 77.5. Over what value on the X-axis should this point appear?

_____13. In your cumulative percentage frequency curve, there should be a point representing a cumulative percentage of 92.5. Over what value on the X-axis should this point appear?

Using the technique described in Section 3.6 of the text, read the value of the following quantities from your ogive.

_____ 14. C_{40}

_____ 15. C_{90}

_____ 16. The centile rank of a clear-day count of 75

_____ 17. The centile rank of a clear-day count of 135

If you won't get these pages back before you do the homework for Chs. 4 and 5, keep a copy of the graphs.

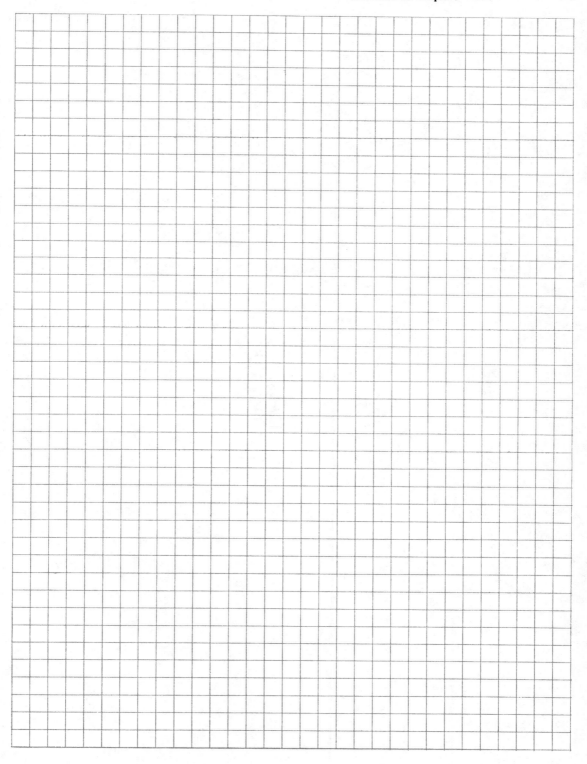

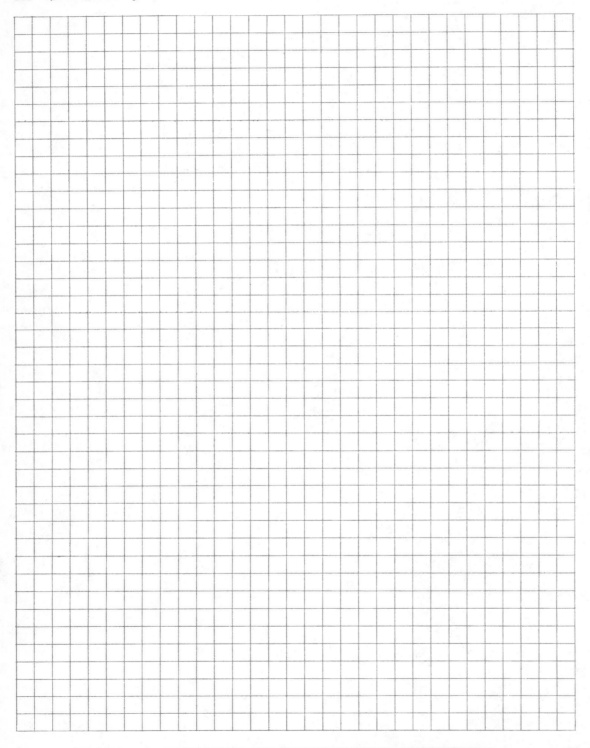

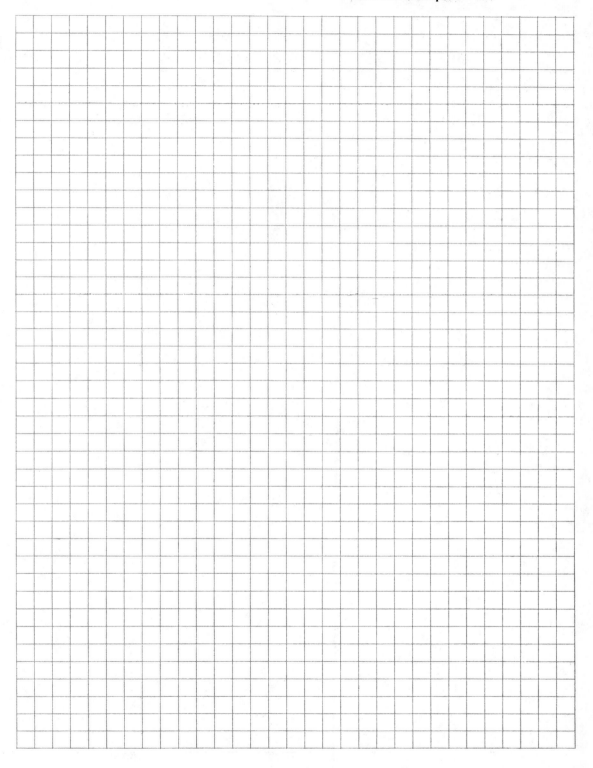

HOMEWORK FOR CHAPTER 4

Look back at the list of 40 clear-day counts in the appendix. Leaving the scores *un*grouped, determine the value of each measure of central tendency.

_____ 1. The mode.

_____ 2. The median. Figure it using the informal procedure described in the sentence spanning pp. 53-54.

_____ 3. The mean. Give it to one decimal place, showing your computation below.

As you have no doubt observed, Phoenix and Sacramento enjoy many, many more sunny days than the other cities for which the appendix reports data. The clear-day counts for these two locations are thus highly discrepant from the remainder of the data, and the issue arises of how much these two extreme scores have affected the three measures of central tendency. Omit these two scores from the distribution, and for the remaining 38 (leaving them ungrouped), determine the mode, median, and mean.

_____ 4. The mode of the 38 scores.

_____ 5. The median of the 38 scores (again determined via the informal procedure described in the sentence spanning pp. 53-54).

_____ 6. The mean of the 38 scores. Again give it to one decimal place, showing your computation below.

For your own edification and amusement, compare each measure of central tendency for the 38 scores with its counterpart for all 40 scores, and note how the information in the last paragraph of p. 59 of the text applies here.

Now work with the grouped frequency distribution that you constructed for Ch. 2's homework. (The 40 scores are grouped into intervals 10 scores wide, beginning with 50-59.) For the distribution grouped in this way, find the mode, the median, and the mean.

_____ 7. The mode of the 40 scores as grouped. (See Exercise 6 for Chapter 4 for the principle by which the mode is to be calculated.)

_____ 8. The median of the grouped distribution. Use the formal procedure, and report the median to one decimal place, showing your computation below. You can check your answer by reading C_{50} from the cumulative percentage frequency curve that you constructed for Ch. 3's homework.

_____ 9. The mean of the grouped distribution. (See Exercise 9 for Chapter 4 for the principle by which the mean is to be calculated.) Give the mean to one decimal place, and show your computation below.

Comparing a measure of central tendency for the grouped data with its counterpart for the ungrouped data will illustrate for you some examples of grouping error.

The following questions concern a distribution of 10 scores. Nine of them are: 3, 5, 9, 1, 9, 2, 0, 3, and 9. The tenth score (the mystery score) is greater than 5 and not 9.

_____ 10. On the basis of the information above, is it possible to determine the mode of the distribution of 10 scores? If yes, write the mode in on the line to the left of the question number. If no, say below why it cannot be determined.

_____ 11. On the basis of the information above, is it possible to determine the median of the distribution of 10 scores by the informal procedure explained in the sentence spanning pp. 53-54? If yes, write the median in on the line to the left of the question number. If no, say below why it cannot be determined.

_____ 12. On the basis of the information above, is it possible to determine the mean of the distribution of 10 scores? If yes, write the mean in at the left. If no, say below why it cannot be determined.

_____ 13. It is possible to change one of the known scores in such a way that the present mode changes. What is the score that when changed has this effect?

_____ 14. To what value must this score be changed to create a new mode that is the only mode of the distribution?

_____ 15. What is the new mode?

_____ 16. Now return to the original distribution. There is one or more scores in the original distribution for which the following assertion is true: if this score changes to the value 5, and no other change in the distribution occurs, the median of the 10 scores will change. (Again, the median is to be calculated by the informal procedure explained on pp. 53-54.) For which score or scores (among the known values) is that assertion true? Write the score or scores in at the left.

_____ 17. In the case or cases that you have just identified, what will the median become?

_____ 18. Again return to the original distribution. There is one or more scores in the original distribution for which the following assertion is true: if this score changes to the value 3, and no other change in the distribution occurs, the median of the 10 scores will change. (The median is to be calculated informally.) For which score or scores is that assertion true? Write the score or scores in at the left.

_____ 19. In the case or cases that you have just identified, what will the median become?

_____ 20. Once more return to the original distribution. There is one or more known scores in the original distribution for which the following assertion is true: if this score changes in *any* way, and no other change in the distribution occurs, the mean of the distribution will change. For which score or scores is that assertion true? Write the score or scores in at the left.

_____ 21. If the mean of the original distribution is 4.8, what is the value of the mystery score? State its value at the left, and show below how you calculated it.

Assume the mean of the original distribution of 10 scores is 4.8. For each change in the distribution described below, say whether the mean would increase ($+$), remain unchanged (0), or decrease ($-$).

$+$ 0 $-$ 22. A value of 6 becomes the eleventh score.

$+$ 0 $-$ 23. A value of 4.8 becomes the eleventh score.

$+$ 0 $-$ 24. A value of -19 becomes the eleventh score.

$+$ 0 $-$ 25. The present value 2 is removed, leaving nine scores.

$+$ 0 $-$ 26. The mystery value is removed, leaving nine scores. (Remember what was stated above about the mystery score.)

HOMEWORK FOR CHAPTER 5

Again turn back to the list of 40 clear-day counts in the appendix to this *Guide*. Leaving them *un*grouped, calculate the following quantities, which include the measures of variability for the distribution. Write your answers in at the left of the questions, and show your work to the right and below a given question.

_____ 1. The range.

_____ 2. Q_1. In accord with the method for ordered raw scores described on p. 68 (the method analgous to the informal procedure for finding the median described in the sentence spanning pp. 53-54), take Q_1 to be midway between the tenth and the eleventh scores (when the scores are arranged in order from low to high).

_____ 3. Q_3. Take Q_3 to be midway between the thirtieth and the thirty-first scores.

_____ 4. Q to one decimal place.

_____ 5. ΣX^2. There is no need to show your work for this quantity or for the next two.

_____ 6. ΣX.

_____ 7. $(\Sigma X)^2$.

_____ 8. Σx^2 to two decimal places.

_____ 9. S^2 to one decimal place.

_____ 10. S to one decimal place.

To check whether your values for S^2 and S are reasonable, you can use the facts about the quantities presented in that section of Ch. 5 of this *Guide* that provides special help with them.

As in calculating the measures of central tendency, the question arises of what those two extreme cases, Phoenix and Sacramento, are doing to the measures of variability. To find out, omit their scores (190 and 201) and recalculate the various quantities you just found. The distribution you are working with now has 38 scores.

_____ 11. The range.

_____ 12. Q_1. Take it to be the tenth score.

_____ 13. Q_3. Take it to be the twenty-ninth score.

_____ 14. Q to one decimal place.

_____ 15. ΣX^2.

_____ 16. ΣX.

_____ 17. $(\Sigma X)^2$.

_____ 18. Σx^2 to two decimal places.

_____ 19. S^2 to one decimal place.

_____ 20. S to one decimal place.

_____ 21. Which of the four measures of variability is least affected by the two extreme scores? (Check to see whether the facts about this particular distribution accord with the generalizations on p. 75 about the effect of extreme scores.)

If you're curious about the measures of variability for the grouped presentation of the 40 clear-day counts, here they are: $Q = 17.5$ ($Q_1 = 79.5$; $Q_3 = 114.5$), $S^2 = 1153.4$, and $S = 34.0$. (If you wish to know how to calculate these values, see Exercise 15 for Ch. 5.) The discrepancy between a value for the grouped data and the counterpart for the ungrouped scores illustrates grouping error.

Here is a distribution of scores created to yield integer values for the mean, the variance, and the standard deviation. Calculate these quantities using the deviation-score method illustrated in Table 5.1 of the text (and in the exercises in Ch. 5 of this *Guide*). Show your work in and below the table.

X	x	x^2
24		
23		
20		
20		
18		
15		

_____ 22. $\bar{X}$ to one decimal place.

_____ 23. S^2 to one decimal place.

_____ 24. S to one decimal place.

25. Now change the distribution so that each score deviates from the mean twice as much as it did before (but in the same direction). This change leaves the mean unaltered but expands the scores outward from it: those above the mean become twice as far above it, and those below become twice as far below. (Any score at the mean stays at the mean, for twice a deviation of zero is still zero.) Display the new distribution below in the column headed X.

X	x	x^2

_____ 26. What is S^2 for the new distribution?

_____ 27. What is the ratio of the new variance to the old?

_____ 28. What is S for the new distribution?

_____ 29. What is the ratio of the new standard deviation to the old?

If you did Questions 22-29 correctly, you have illustrated a generalization about the variance and the standard deviation, and you can probably guess what the generalization is. It appears below with two words missing; fill them in.

If the variability in a distribution increases such that each score becomes twice as far from the mean, the standard deviation becomes _____ times as great, and the variance becomes _____ times as great.

HOMEWORK FOR CHAPTER 6

These exercises will probably strike you as repetitious and tedious. Please do them carefully, though, because the principles you'll be learning and practicing are essential for the understanding of important and interesting matters coming up in this course.

Answer these questions to four decimal places (e.g., as .1234). If a collection of scores is normally distributed, what proportion fall ...

_____ 1. above $z = +0.50$?   _____ 2. above $z = +1.50$?

_____ 3. below $z = -2.50$?   _____ 4. below $z = -3.50$?

_____ 5. above $z = -0.50$?   _____ 6. above $z = -1.50$?

_____ 7. below $z = +2.50$?   _____ 8. below $z = +3.50$?

_____ 9. between $z = +0.60$ and $z = +1.20$?

_____ 10. between $z = -1.80$ and $z = -2.40$?

_____ 11. between $z = -1.80$ and $z = +0.60$?

_____ 12. between $z = -2.40$ and $z = +1.20$?

_____ 13. outside the limits $z = +0.40$ and $z = +0.80$?

_____ 14. outside the limits $z = -1.20$ and $z = -1.60$?

_____ 15. outside the limits $z = -0.40$ and $z = +0.40$?

_____ 16. outside the limits $z = -0.80$ and $z = +0.80$?

IQ scores for the general public on the Wechsler Adult Intelligence Scale (the WAIS) are normally distributed with a mean of 100 and a standard deviation of 15. Answer the following questions to two decimal places (e.g., as 12.34%). What percentage of the general (adult) public has a Wechsler IQ ...

_____ 17. below 70?   _____ 18. below 85?

_____ 19. above 115?   _____ 20. above 145?

_____ 21. below 130?   _____ 22. above 85?

_____ 23. between 85 and 115?   _____ 24. between 70 and 130?

_____ 25. between 55 and 145?

Answer the following questions to two decimal places (e.g., as 1.23). If a collection of scores is normally distributed, what z score ...

_____ 26. divides the upper 10% of the scores from the remainder?

_____ 27. divides the upper 20% of the scores from the remainder?

_____ 28. divides the lower 25% of the scores from the remainder?

_____ 29. divides the lower 40% of the scores from the remainder?

_____ 30. divides the upper 60% of the scores from the remainder?

_____ 31. divides the upper 70% of the scores from the remainder?

_____ 32. divides the lower 80% of the scores from the remainder?

_____ 33. divides the lower 90% of the scores from the remainder?

Answer the questions below to two decimal places again. If a collection of scores is normally distributed, what z score limits identify ...

_____ 34. the central 90% of the scores?

_____ 35. the central 80% of the scores?

_____ 36. the outermost 70% of the scores?

_____ 37. the outermost 60% of the scores?

Answer these questions to the nearest whole number (e.g., as 123). If a distribution of scores on a standardized aptitude test is normal in shape with a mean of 500 and a standard deviation of 100, what is the raw score (not the z score) ...

_____ 38. below which 50% of the scores fall?

_____ 39. below which 75% of the scores fall?

_____ 40. above which 85% of the scores fall?

_____ 41. above which 95% of the scores fall?

Again answer to the nearest whole number. In the distribution described just above, what are the raw scores (not the z scores) ...

_____ 42. that enclose the central 50% of the scores?

HOMEWORK FOR CHAPTER 7

1. In the appendix to this *Guide* there appear some data on the students who took the statistics course offered in the fall of 1977 at Ramapo College of New Jersey. For those students, how closely was their performance on the first test in the course related to their performance over the entire semester? To begin to answer this question, make a scatter diagram of the data on the back of this sheet. Do it neatly and as large as possible.

To provide a more precise answer to the question about the relationship between performance on the first exam and performance over the entire semester, compute the Pearson correlation coefficient for the data. Use the raw-score method illustrated on p. 121 of the text, and find the means and standard deviations of both variables while you're at it.

Don't bother to list the individual values of X^2 and Y^2, but do show your other work. Give all values that are not whole numbers to three decimal places (as 1.234, e.g.). This is more than the number that would usually be reported for the mean or the standard deviation or a correlation coefficient, but you'll need the extra accuracy for future work with these data.

2. ΣX: _____   7. ΣY: _____

3. ΣX^2: _____   8. ΣY^2: _____

4. Σx^2: _____   9. Σy^2: _____

5. $\bar{X}$: _____   10. $\bar{Y}$: _____

6. S_X: _____   11. S_Y: _____

12. ΣXY: _____   13. Σxy: _____

14. r: _____

Now that you've found the means, go back to your scatter diagram and add lines that show the locations of the means, as in the figure on p. 118 of your text.

_____ 15. Which students' data points lie in the first quadrant? List the letters that identify the students.

_____ 16. Which students' data points lie in the second quadrant?

_____ 17. Which students' data points lie in the third quadrant?

_____ 18. Which lie in the fourth quadrant?

_____ 19. What proportion of the variance in the percentage correct over the entire semester is associated with variation in the percentage correct on the first exam? Answer to two decimal places.

If you won't get this page back before you do the homework for the next chapter, keep a copy of all your answers.

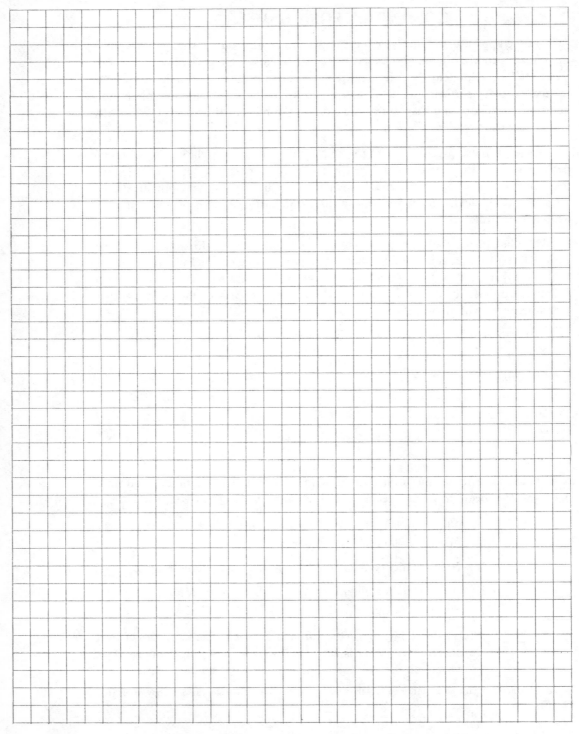

HOMEWORK FOR CHAPTER 8

Turn back to the appendix in this *Guide* and again inspect the data on the students in the statistics course offered at Ramapo College of New Jersey in 1977. Use those data to determine the regression equation for predicting percentage correct over the entire semester from percentage correct on the first exam. For Formula 8.1 in the text, you'll need the means and standard deviations of the two variables and the correlation coefficient, which you computed in the homework for Ch. 7. Use all three of the decimal places to which you were asked to give those numbers, and round the figures in the regression equation to two decimal places. Simplify the equation as much as possible, until it has the form of the equation in the second line on p. 140 of the text. Write the simplified equation in below, and show your work below it.

$$Y' = \text{\underline{\hspace{5cm}}}$$

2. Now make another scatter diagram of the data, as you did for Ch. 7, but this time add the regression line to it. Use the back of this sheet.

_____ 3. The student who was at the median on the first exam (and who was omitted from the data you've working with) earned a score of 69% correct on that exam. Use your regression equation to predict this person's score over the entire semester (to two decimal places). The actual figure was 86% correct. Show your work below.

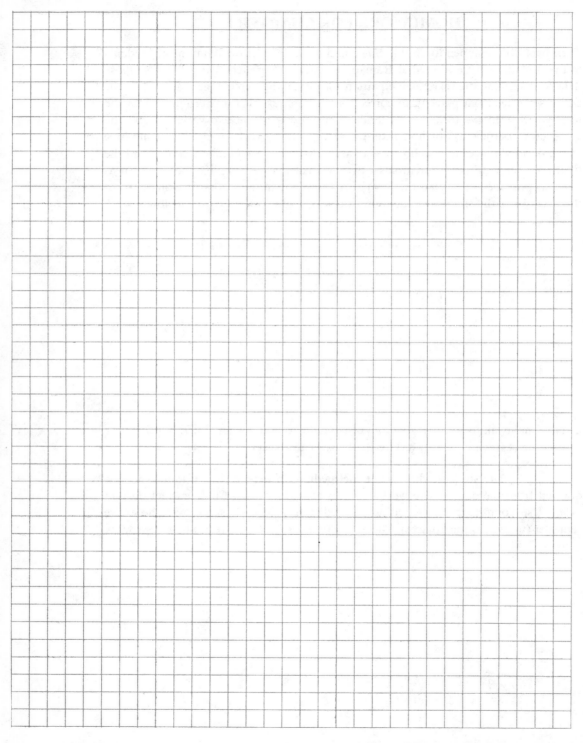

4. Now use the regression equation to predict performance over the entire semester for the 10 students who contributed data to the table. Fill in the table below. Give the values of Y' to two decimal places. $\Sigma(Y - Y')^2$ should be zero, but it may be off a little because of rounding error.

Student	X	Y	Y'	$(Y-Y')$	$(Y-Y')^2$
A	38	75			
B	54	65			
C	62	94			
D	67	81			
E	67	84			
F	72	93			
G	77	90			
H	77	93			
I	82	90			
J	85	95			

_____ 5. Compute S_{YX} to two decimal places, doing it directly as $\sqrt{\Sigma(Y-Y')^2/n}$, in accord with Formula 8.3. Show your work.

_____ 6. Now compute S_{YX} to two decimal places from Formula 8.4, again showing your work. If you didn't get the same value, something is wrong.

HOMEWORK FOR CHAPTER 9

An honest die is one whose six faces turn up with equal probability. The principle in the second box on p. 156 thus applies. The questions below refer to a pair of dice, and you should assume that both are honest. In answering the questions, give the requested probabilities *both* as common fractions reduced as far as possible (e.g, as 1/6) *and* as decimal fractions to four decimal places (as .1234, e.g.). You'll do best if you first translate each question into an OR question, an AND question, or a combination of the two, whichever is appropriate.

With a pair of honest dice, what is probability of rolling ...

_____ 1. a 5 on a certain one of the dice (call it Die A) and a 6 on the other (on Die B)?

_____ 2. a 6 on Die A and a 5 on Die B?

_____ 3. a 5 on one die (it doesn't matter which) and a 6 on the other?

_____ 4. a 6 on one die (it doesn't matter which) and a 6 on the other?

_____ 5. "doubles" (the same number on both dice)?

To draw a sample of some given size at random is to draw it in such a way that all possible samples of that size are equally likely. In drawing a card (a sample of size 1) at random from a deck of playing cards, then, the principle in the second box on p. 156 of the text applies. Answer the following questions as you did the series above.

If you draw a card at random from a standard deck of 52, what is the probability that you will get ...

_____ 6. the Queen of Hearts?

_____ 7. a queen (of any kind)?

_____ 8. a heart (of any kind)?

_____ 9. a queen OR a heart?

Now suppose you draw a first card at random, look at it, replace it, and draw again at random. (This is equivalent to drawing the first card from one deck and the second card from a second deck.)

What is the probability that you will get ...

_____ 10. the Queen of Hearts on both draws?

_____ 11. a queen (of any kind) on both draws?

_____ 12. a heart (of any kind) on both draws?

_____ 13. a queen on the first draw AND a heart on the second?

_____ 14. The LaMaze method is a technique of prepared childbirth permitting a laboring woman to participate actively in the delivery of her child, perhaps obviating the need for analgesia or anesthesia. In the LaMaze class attended by the sister of the author of this *Guide*, there were seven women enrolled, and all of them gave birth to a girl rather than a boy. Is this a rare occurrence? Assume that the probability of any one woman's bearing a girl is 1/2. (Actually, the probability of a boy is slightly greater than the probability of a girl, about .51 or .52. More males than females are conceived, and even though male fetuses are more likely to die during gestation, they still predominate slightly at delivery time.) Find the probability that seven women bearing children (with no multiple births) will each deliver a girl. Show your work at left.

_____ 15. What is the probability that those seven women will all have a child of the same sex?

16. In answering Questions 14 and 15, what assumption (other than that Pr(girl)= Pr(boy)=1/2) did you have to make?

_____

HOMEWORK FOR CHAPTER 10

Suppose you're conducting research on something like errors in social judgment, something that might be influenced by your subjects' intelligence. You're accordingly worried about getting a sample of subjects who, as a group, are unusually bright or unusually dull. You know that IQ scores for the general adult public on the Wechsler Adult Intelligence Scale (the WAIS) are normally distributed with a mean of 100 and a standard deviation of 15. If your subjects will be a truly random sample of this population, you can correctly predict the likelihood of getting a group whose mean IQ lies more than, say, 5 or 10 points away from the population mean of 100. So, onward to the problems below, the answers to which will tell you whether it's realistic to worry about such things as getting a sample whose mean IQ is below 90 or over 110. Answer the questions to four decimal places (as .1234, e.g.), and show your work.

First, suppose your sample size is going to be nine. What is the probability that those nine people will have IQ scores whose mean is ...

_____ 1. over 105?

_____ 2. under 90?

_____ 3. more than 5 points away from 100 in either direction (that is, more than 100+5 or less than 100−5)?

_____ 4. more than half a standard deviation away from the population mean? (The standard deviation in an expression like this is that of the population of raw scores—here, IQ scores.)

_____ 5. *within* one standard deviation of the population mean?

Now suppose you almost triple your sample size, increasing it to 25. Again answer to four decimal places, and show your work. What is the probability that the 25 people in your sample will have IQs with a mean ...

_____   6. over 105?

_____   7. under 90?

_____   8. more than 5 points away from the population mean?

_____   9. more than half a standard deviation away from the population mean?

_____   10. within one standard deviation of the population mean?

_____   11. Suppose your sample will be quite large, with 100 persons. To four decimal places, what is the probability that the 100 people will have Wechsler IQs whose mean lies within one measly point of the population mean? You may find the answer surprisingly large.

_____   12. With a sample as large as 100, it doesn't much matter whether the distribution of IQs is normal in the population. Even if it departs considerably from normality, we can still be quite confident that the answer to Question 11 is correct. Why is this? There are three "magic" words that name the reason, and the briefest possible answer to this question (an answer that is still entirely correct, though) requires no more than those three little words.

HOMEWORK FOR CHAPTER 11

Under the "personalized system of instruction" (PSI), material to be learned is divided into small units, students work on it at their own pace, and they take the exam on a given unit only when they think they're ready for it. They must pass the exam at a high level before going ahead to the next unit, but they are allowed several tries for each unit (taking a different exam each time, of course).

Three psychologists at Southwest Minnesota State College have reported a comparison of PSI with the traditional mode of instruction for introductory psychology (R. C. Riedel, B. Harney, & W. LaFief, "Unit Test Scores in PSI versus Traditional Classes in Beginning Psychology," *Teaching of Psychology*, 1976, *3*, 76-78). In the fall quarter of an academic year, they used the PSI method with a criterion of 16 out of 20 correct (80%) as the passing score for the test on each of the ten units into which they divided their course. They did not lecture but merely made themselves available at certain times to administer the exams to any students who were seeking to take one. In the winter quarter the psychologists did lecture, and they administered the same tests, one for each unit, every fourth class period, with no opportunity for students to retake an exam.

The psychologists expected that the students in the PSI course would generally make their initial try at the exam on a given unit without being fully prepared, so that most would not meet the criterion of 16 correct on the first try. In fact, in the students' first tries at the exam on the very first unit of the course, they earned a mean of 17.93 correct (which is almost 90%). The psychologists reported the n and s: 64 and 2.93, respectively. Thus you can determine whether it is plausible that those 64 scores are a random sample from a population whose mean is only 16. Test the appropriate hypothesis, using the .05 level of significance and doing a nondirectional test. Carry all calculations to three decimal places, and round the answers you report to two—but if you use an answer in a later calculation, use the three-place version.

1. H_0 in symbols: _____   2. H_A in symbols: _____

3. α: _____   4. $\bar{X}$: _____   5. $s_{\bar{X}}$: _____

6. Critical value(s) of t (use the normal curve): _____

7. t: _____   8. Decision on H_0:   Accept   Reject

9. What does the statistical decision in Question 8 mean in substantive terms? That is, what are its implications for the substantive question of whether the PSI students were generally unprepared to meet the criterion of 16 correct on their first tries at the exam?

Yes   No 10. If you had conducted that test at the .01 level of significance, would your decision on the null hypothesis have been different?

Yes   No 11. If you had conducted the test at the .05 level of significance but had done a one-tailed version in which the alternative hypothesis stated that $\mu < 16$, would your decision on the null hypothesis have been different?

Yes   No 12. If you had conducted the test at the .01 level of significance and had done a one-tailed version in which the alternative hypothesis stated that $\mu < 16$, would your decision on the null hypothesis have been different?

In the lecture course, the mean score on the first exam was only 13.41 (67% correct). Is it plausible that the scores in this group are a random sample from a population whose mean is as large as 16? The n for the group was 61, and s was 4.10. Do a two-tailed test at the .01 level of significance.

13. H_0 in symbols: _____   14. H_A in symbols: _____

15. α: _____   16. $\overline{X}$: _____   17. $s_{\overline{X}}$: _____

18. Critical value(s) of t (use the normal curve): _____

19. t: _____   20. Decision on H_0:   Accept   Reject

Over the remaining nine units of the course, the PSI students earned a mean above 16 on their first tries at every unit except one. The mean of their first tries at the eighth unit was only 15.39 ($n=62$, $s=2.80$). Is it plausible that this time the scores were a random sample from a population whose mean was only 16? Do the appropriate two-tailed test at the .05 level of significance.

21. H_0 in symbols: _____   22. H_A in symbols: _____

23. α: _____   24. $\overline{X}$: _____   25. $s_{\overline{X}}$: _____

26. Critical value(s) of t (use the normal curve): _____

27. t: _____   28. Decision on H_0:   Accept   Reject

_____29. If you had conducted the test (still two-tailed) at the .10 level of significance, what would the critical value(s) have been? (Use the normal curve.)

Yes   No 30. Would your decision on H_0 have been different?

You may be interested to know that over all ten units of the course, even on their first tries, the PSI students earned a mean score higher than the mean for the students in the lecture version. In the homework for Ch. 13, you will have a chance to determine whether the differences are statistically significant.

HOMEWORK FOR CHAPTER 12

Turn back to the appendix of this *Guide* and inspect the column of data on the performance of those statistics students on the first examination in their course. The figure 81.25% is of some interest here, because it is the score we would expect a student to earn if she or he knows 75% of the questions and has average luck in guessing at the remainder. Is it plausible that those ten scores are a random sample from a population whose mean is 81.25? Presume you are interested to know both whether the population mean is below 81.25 and whether it is above 81.25. Test the appropriate hypothesis, using the .05 level of significance. Carry all your calculations to three decimal places, round the answers to two in reporting a value, but use the three-place version if you need the value in a later computation. (Some of the quantities requested below you already calculated in the homework for Ch. 7.)

1. H_0 in symbols: _____   2. H_A in symbols: _____

3. α: _____   4. $\bar{X}$: _____

5. ΣX^2: _____   6. ΣX: _____

7. $(\Sigma X)^2$: _____   8. Σx^2: _____

9. s: _____   10. $s_{\bar{X}}$: _____

11. df: _____   12. t_{crit}: _____

13. t_{calc}: _____   14. Decision on H_0: Accept   Reject

Now test the same hypothesis in the same way for the data on the students' performances over the entire semester. (These are thescores you previously called Y.)

15. H_0 in symbols: _____   16. H_A in symbols: _____

17. α: _____   18. $\bar{X}$: _____

19. ΣX^2: _____   20. ΣX: _____

21. $(\Sigma X)^2$: _____   22. Σx^2: _____

23. s: _____   24. $s_{\bar{X}}$: _____

25. df: _____   26. t_{crit}: _____

27. t_{calc}: _____   28. Decision on H_0: Accept   Reject

    You may be wondering whether the mean level of performance improved in the population that those ten students represent. And you may be tempted to draw a conclusion on this issue by comparing the outcomes of the two hypothesis tests you just did. Such a procedure is crude, indirect, and sometimes erroneous. It is better to compare the sample means directly, in a single test, a test of the hypothesis that the two populations have the same mean. Chs. 13 and 14 will show you how.

Suppose a researcher does a two-tailed test of a null hypothesis stating that the mean of a certain population is 500. The mean may be less than 500, exactly 500, or greater than 500. The researcher could accept or reject the null hypothesis. In all, there are six possible combinations, and these are listed below, one possible combination per column.

In each column, indicate whether the researcher's decision is correct (C), a Type I error (I), or a Type II error (II).

	29.	30.	31.	32.	33.	34.
In reality:	$\mu = 500$	$\mu = 500$	$\mu < 500$	$\mu < 500$	$\mu > 500$	$\mu > 500$
Decision on H_0:	Accept	Reject	Accept	Reject	Accept	Reject
	C	C	C	C	C	C
	I	I	I	I	I	I
	II	II	II	II	II	II

HOMEWORK FOR CHAPTER 13

In the homework for Ch. 11, you learned of a study comparing PSI and a conventional lecture course in elementary psychology. The instructors of the courses reported the following data for scores on the first exam:

	PSI	Lecture
$\overline{X}$	17.93	13.41
s	2.93	4.10
n	64	61

Is it plausible that the two classes are random samples from populations with identical means? Test the appropriate hypothesis at the .01 level of significance, doing a two-tailed test.

    1. These samples are (circle one): independent   dependent

2. H_0 in symbols: _____

3. H_A in symbols: _____

4. α: _____

5. $\overline{X}_1 - \overline{X}_2$: _____

6. s_p^2* to three decimal places: _____

7. $s_{\overline{X}_1 - \overline{X}_2}$ to three decimal places: ___

8. t_{calc} to two decimal places: _____

9. df: _____

10. t_{crit}: _____

11. Decision on H_0: Accept   Reject

12. What are the implications of your statistical conclusion in Question 11 for the substantive question of whether the PSI students' first tries at the test would not be as good as the lecture students' performance?

    The mean for the PSI students' first attempts at each of the remaining nine exams was higher than the corresponding mean for the lecture students (even though the latter knew they had only one chance at an exam). Of these nine units, two are of special interest. The third unit is of interest, because that was the one on which the lecture students did the best and the only unit on which they earned a mean over 16. The data are as follows:

	PSI	Lecture
$\overline{X}$	17.06	16.60
s	2.03	2.91
n	64	72

* To calculate s_p^2 from s_1^2 and s_2^2, see Footnote 2 on p. 233 of the text.

Again, determine whether it is plausible that the parent populations have the same mean, doing a two-tailed test at the .01 level of significance.

13. These samples are (circle one): independent   dependent

14. H_0 in symbols: _____   15. H_A in symbols: _____

16. α: _____   17. $\bar{X}_1 - \bar{X}_2$: _____

18. s_p^2 to three decimal places: _____   19. $s_{\bar{X}_1 - \bar{X}_2}$ to three decimal places: ___

20. t_{calc} to two decimal places: _____   21. df: _____

22. t_{crit}: _____   23. Decision on H_0: Accept   Reject

Also of interest are the data on the eighth unit, because this was the one on which the PSI students did least well, and the only one on which the mean of their first tries at the exam was under 16. The PSI class still outperformed the lecture class, though. Do a two-tailed test at the .01 level of significance again. Here are the data:

	PSI	Lecture
$\bar{X}$	15.39	14.27
s	2.80	3.37
n	62	61

24. These samples are (circle one): independent   dependent

25. $\bar{X}_1 - \bar{X}_2$: _____   26. s_p^2 to three decimal places: _____

27. $s_{\bar{X}_1 - \bar{X}_2}$ to three decimal places: ___   28. t_{calc} to two decimal places: _____

29. df: _____   30. t_{crit}: _____

31. Decision on H_0: Accept   Reject

None   1st   2nd   3rd 32. Which, if any, of the three tests you've just done would have yielded a different conclusion about the null hypothesis if it had been conducted at the .05 level of significance?

33. Suppose you're wondering whether the PSI students' performance on their first tries on the first unit differed significantly from their performance on their tries at the tenth unit ($\bar{X} = 16.49$). If you were to do a two-tailed test of the hypothesis that the parent populations for the two samples of scores had identical means, using the .01 level of significance again, could you follow the procedure and use the formulas that you employed for the three problems above? Circle your answer and justify it: Yes   No

HOMEWORK FOR CHAPTER 14

Here are the data on the examination performance of those ten statistics students whom you met in the homework for Chs. 7 and 8.

Student	X_1	X_2	X_1^2	X_2^2	$X_1 X_2$
A	38	75			
B	54	65			
C	62	94			
D	67	81			
E	67	84			
F	72	93			
G	77	90			
H	77	93			
I	82	90			
J	85	95			

Did the students' total-performance percentages significantly exceed their percentage performance on the first exam? Test the difference between the two means using the method illustrated in Table 14.1. You have already done most of the necessary calculations in the homework for Ch. 7. In constructing the alternative hypothesis, assume that you are interested in discovering only whether the total-performance mean exceeds the first-exam mean in the populations. Use the .01 level of significance. Carry your calculations to at least three decimal places, and round to two places in reporting an answer—but if you use a value in a later computation, use the three-place version.

1. These samples are: independent dependent

2. H_0 in symbols: _____   3. H_A in symbols: _____

4. α: _____

5. ΣX_1: _____   9. ΣX_2: _____

6. ΣX_1^2: _____   10. ΣX_2^2: _____

7. Σx_1^2: _____   11. Σx_2^2: _____

8. $\overline{X}_1$: _____   12. $\overline{X}_2$: _____

13. $\overline{X}_1 - \overline{X}_2$: _____

14. $\Sigma X_1 X_2$: _____

15. $\Sigma x_1 x_2$: _____

16. $s_{\overline{X}_1 - \overline{X}_2}$: _____

17. t_{calc}: _____

18. df: _____

19. t_{crit}: _____

20. Decision on H_0: Accept Reject

To aid in the interpretation of the data, also calculate the following quantities. Give the s's to just one decimal place.

21. s_1: _____

22. s_2: _____

23. r_{12} (from the homework for Ch. 7): _____

Here are those data again:

Student	X_1	X_2	D	D^2
A	38	75		
B	54	65		
C	62	94		
D	67	81		
E	67	84		
F	72	93		
G	77	90		
H	77	93		
I	82	90		
J	85	95		

Test the hypothesis of no improvement from the first exam to the semester total using the method of difference scores illustrated in Table 14.2 of the text. Again, assume you're interested only in whether the total-performance mean exceeds the first-exam mean in the populations. And again, work at the .01 level of significance. Start by listing the difference scores in the table. Calculate each as the performance on the first exam minus the total performance (as $D = X_1 - X_2$).

24. H_0 in terms of D: _____

25. H_A in terms of D: _____

26. α: _____

27. ΣD: _____

28. $\overline{D}$: _____

29. ΣD^2: _____

30. Σd^2: _____

31. s_D: _____

32. $s_{\overline{D}}$: _____

33. t_{calc}: _____

34. df: _____

35. t_{crit}: _____

36.   Decision on H_0:   Accept   Reject

Did you get the same value for t this time? And remember that when the arithmetic is correct, $\overline{D} = \overline{X}_1 - \overline{X}_2$ and $s_{\overline{X}_1 - \overline{X}_2} = s_{\overline{D}}$.

HOMEWORK FOR CHAPTER 15

In a study conducted at a street corner with a traffic light, an experimenter stood on the curb waiting for the light to turn from red to green. When another pedestrian walked up, the experimenter turned and gave either a quick glance or a prolonged stare. Regardless of the sex of the experimenter or the sex of the subject, those who had been stared at tended to cross the street faster, once the light changed, than those who had received only a glance. The mean crossing time for 66 subjects in the stare condition was 11.1 seconds, while the mean for 62 subjects in the glance condition was 12.2 seconds, and this difference proved to be highly significant. The standard deviation of the population of crossing times was estimated to be 1.00 seconds (really) in both conditions.*

It seems safe to assume that the researchers conducted a two-tailed test of the null hypothesis that the population means were equal, and that they used the .05 level of significance. One wonders how small a difference between the population means they could have discovered in this procedure with samples of the sizes they employed. There is no one answer, of course; rather the smaller the difference, the lower the probability of their discovering it. But if the probability of discovery is specified, one can estimate the difference whose discovery carried this probability. Do so for the probabilities named below, using Figure 15.3. (Only approximate answers will be possible, but do your best to get d to two decimal places.)

_____ 1. What value of d was discoverable with a probability of .80?

_____ 2. To what difference between the population means does this value of d correspond?

_____ 3. What is the power of the test in this case?

_____ 4. What is the risk of wrongly accepting the null hypothesis in this case?

_____ 5. What value of d was discoverable with a probability of .90?

_____ 6. To what difference between the population means does this value of d correspond?

_____ 7. What is the power of the test in this case?

---

* This study was the work of P. C. Ellsworth, J. M. Carlsmith, and A. Henson, who reported it in their article "The Stare as a Stimulus to Flight in Human Subjects" in the *Journal of Personality and Social Psychology*, 1972, *21*, 302–311.

_____ 8. What is the risk of a Type II error in this case?

_____ 9. What value of d was discoverable with a probability of fully .95?

_____ 10. To what difference between the populations means does this value of d correspond?

_____ 11. What is the power of the test in this case?

_____ 12. What is the risk of missing the difference between the population means in this case?

Below is a list of ways in which the researchers could have altered their experimental design. For each way, indicate whether the power of their test would have increased $(+)$, stayed the same (0), or decreased $(-)$.

$+$ 0 $-$ 13. They use 100 subjects in each condition.

$+$ 0 $-$ 14. They use the .01 level of significance.

$+$ 0 $-$ 15. They employ a one-tailed test in which the alternative hypothesis states that $\mu_{stare} > \mu_{glance}$. (Assume that the stare speeds up crossing time, as the results strongly suggest.)

$+$ 0 $-$ 16. They use the .10 level of significance.

$+$ 0 $-$ 17. They use 20 subjects in each condition.

$+$ 0 $-$ 18. They employ a one-tailed test in which the alternative hypothesis states that $\mu_{stare} < \mu_{glance}$. (Assume again that the stare speeds up crossing time.)

HOMEWORK FOR CHAPTER 16

In doing the homework for Chs. 12-14, you employed the inferential technique of hypothesis testing. The problems for this chapter ask you to make interval estimates for the same data. In working the problems, report $s_{\bar{X}}$ or $s_{\bar{X}_1 - \bar{X}_2}$ to two decimal places, but in calculating the limits use the value with three decimal places. Report the limits to two decimal places. Show your work.

On their first tries at their first exam, the 64 PSI students described in the homework for Ch. 11 earned a mean of 17.93 on a 20-item test; s was 2.93. Find the 95% confidence interval for the population mean.

_____ 1. df _____ 2. t_c _____ 3. $s_{\bar{X}}$

_____ 4. Lower limit _____ 5. Upper limit

You may wish to compare your interval with the conclusion to the hypothesis test on the same data.

On their first tries on the first exam, the 61 students in the lecture course earned a mean of only 13.41; s was 4.10. Find the 99% confidence interval for the population mean.

_____ 6. df _____ 7. t_c _____ 8. $s_{\bar{X}}$

_____ 9. Lower limit _____ 10. Upper limit

Again, you may wish to compare the confidence interval with the results of the hypothesis test.

Over the ten examinations, the lowest mean for the PSI students was 15.39. This time $n = 62$ and $s = 2.80$. Find the 90% confidence interval for the population mean.

_____ 11. df _____ 12. t_c _____ 13. $s_{\bar{X}}$

_____ 14. Lower limit _____ 15. Upper limit

Now do a direct comparison of the mean on the first test for the PSI students and the corresponding mean for the lecture students. Find the 99% confidence interval for the difference between the two population means.

Independent   Dependent 16. Are the samples independent or dependent?

_____ 17. $\bar{X}_1 - \bar{X}_2$ _____ 18. df _____ 19. t_c

_____ 20. $s_{\bar{X}_1 - \bar{X}_2}$ (see Questions 6 and 7 for Ch. 13)

_____ 21. Lower limit _____ 22. Upper limit

You may wish to compare the confidence interval with the results of the hypothesis test in the homework for Ch. 13.

In the comparison between PSI and lecture students, the third exam was of interest, because that was the one on which the lecture class did the best, and the only one on which they earned a mean over 16. The data were: $\overline{X}_1 = 17.06$ with $s = 2.03$ for 64 PSI students; $\overline{X}_2 = 16.60$ with $s = 2.91$ for 72 lecture students. Find the 95% confidence interval for the difference between the population means.

Independent   Dependent 23. Are the samples independent or dependent?

_____ 24. $\overline{X}_1 - \overline{X}_2$ _____ 25. df _____ 26. t_c

_____ 27. $s_{\overline{X}_1 - \overline{X}_2}$ (see Questions 6 and 7 for Ch. 13)

_____ 28. Lower limit   _____ 29. Upper limit

The comparable hypothesis test is in the homework for Ch. 13.

Also of interest are the data on the eighth unit, because this was the one on which the PSI students did least well, and the only one on which the mean of their first tries fell under 16. The data were: $\overline{X}_1 = 15.39$ with $s = 2.80$ for 62 PSI students; $\overline{X}_2 = 14.27$ with $s = 3.37$ for 61 lecture students. Set the 90% confidence interval for the difference between the population means.

Independent   Dependent 30. Are the samples independent or dependent?

_____ 31. $\overline{X}_1 - \overline{X}_2$ _____ 32. df _____ 33. t_c

_____ 34. $s_{\overline{X}_1 - \overline{X}_2}$ (see Questions 6 and 7 for Ch. 13)

_____ 35. Lower limit   _____ 36. Upper limit

The comparable hypothesis test is again in the homework for Ch. 13.

In the homework for Ch. 14, you were concerned with the ten students tested with 500 multiple-choice items in a statistics course. You tested the difference between the mean percentage correct on the first exam and the mean percentage correct over the entire semester. Now construct the 99% confidence interval for the difference between the population means.

Independent   Dependent 37. Are the samples independent or dependent?

_____ 38. $\overline{X}_1 - \overline{X}_2$ _____ 39. df _____ 40. t_c

_____ 41. $s_{\overline{X}_1 - \overline{X}_2}$ (see Questions 6 and 7 for Ch. 13)

_____ 42. Lower limit   _____ 43. Upper limit

A comparison of the confidence interval with the results of the hypothesis test will again be informative.

HOMEWORK FOR CHAPTER 17

In the homework for Ch. 7, you computed the value of r for the correlation between the percentage correct on a first exam in a statistics course and the percentage correct over the entire semester. (Remember, this is not *the* correlation for these two variables. See Section 7.8 on pp. 125-126 of the text.) Now test the significance of that value of r.

_____ 1. The obtained value of r (to two decimal places).

_____ 2. *df*

_____ 3. H_0 for the test of the significance of r

_____ 4. H_A for a two-tailed test

_____ 5. The critical value(s) of r for $\alpha = .05$

Accept   Reject 6. Your conclusion on H_0

If you were to conduct a two-tailed test at $\alpha = .01$, ...

_____ 7. what would the critical value(s) of r be?

Accept   Reject 8. what would you conclude about H_0?

If you were to conduct a one-tailed test at $\alpha = .05$ with an interest in discovering only whether the correlation coefficient in the population is positive, ...

_____ 9. what would H_A state?

_____ 10. what would the critical value(s) of r be?

Accept   Reject 11. would you accept or reject H_0?

_____ 12. Exactly what does it mean to say that the value of r for the data you are working with is "significant"? Explain this statistical terminology.

_____ 13. Using the formula on p. 299, translate the r value for the data on the statistics course into a t. (Use the value of r with three decimal places in the computation.)

Using this t value and Appendix G, evaluate the value of r for these cases:

_____ 14. What is the critical value of t for a two-tailed test of H_0 at $\alpha = .05$?

Yes No 15. Is r significant in this case according to your t-test?

_____ 16. What is the critical value of t for a two-tailed test of H_0 at $\alpha = .01$?

Yes No 17. Is r significant in this case according to your t-test?

_____ 18. What is the critical value of t for a directional test of H_0 at $\alpha = .05$ if you are interested in discovering only whether the population's correlation coefficient is positive?

Yes No 19. Is r significant in this case according to your t-test?

The answers for Questions 15, 17, and 19 should agree with your conclusions in Questions 6, 8, and 11, respectively.

Now use Figure 17.2 to estimate the 95% confidence limits for the value of r you have been working with.

_____ 20. The lower limit

_____ 21. The upper limit

Yes No 22. Are the two limits equidistant from the value of r?

LL UL 23. If no, is the lower limit (LL) or the upper limit (UL) closer?

If a sample of 50 cases had yielded the same value of r, what would the 95% confidence limits have been?

_____ 24. The lower limit

_____ 25. The upper limit

Yes No 26. Are the two limits equidistant from the value of r?

LL UL 27. If no, is the lower limit or the upper limit closer?

10 50 28. In which case is the 95% confidence interval more narrow and thus more precise: when $n = 10$ or when $n = 50$?

HOMEWORK FOR CHAPTER 18

To become comfortable with the one-way analysis of variance, and to gain insight into its workings, it is helpful to do an analysis with very simple numbers. Such numbers are supplied below, and you will find that most of the quantities derived from them (means, sums of squares, and so on) also turn out to be simple. In doing the analysis, you are asked first to use the definitional formulas of Section 18.5 and then to apply the raw-score formulas of 18.7. You should get the same results both times, of course. Begin by filling in the table.

X_D	$X_D - \bar{X}_D$	$(X_D - \bar{X}_D)^2$	X_E	$X_E - \bar{X}_E$	$(X_E - \bar{X}_E)^2$	X_F	$X_F - \bar{X}_F$	$(X_F - \bar{X}_F)^2$
21			16			12		
21			15			10		
20			15			10		
19			15			10		
19			14			8		

Note that the means are widely dispersed (relatively speaking), whereas within each sample the scores cluster tightly about their own mean.

_____ 1. $\bar{X}_D$ _____ 2. $\bar{X}_E$ _____ 3. $\bar{X}_F$

_____ 4. $\Sigma(X_D - \bar{X}_D)^2$ _____ 5. $\Sigma(X_E - \bar{X}_F)^2$ _____ 6. $\Sigma(X_E - \bar{X}_F)^2$

_____ 7. SS_W _____ 8. df_W _____ 9. s_W^2

Now compute SS_W via the raw-score formula on p. 315.

_____ 10. ΣX^2 (where summation is over all scores)

_____ 11. The quantity in square brackets in Formula 19.1

_____ 12. SS_W as computed from Formula 18.1. Compare with the previous result.

Onward to s_A^2. First use the formula on p. 313, showing your work in computing the numerator of the fraction by filling in the table below.

X_D	$\bar{X}_D - \bar{\bar{X}}$	$(\bar{X}_D - \bar{\bar{X}})^2$	X_E	$\bar{X}_E - \bar{\bar{X}}$	$(\bar{X}_E - \bar{\bar{X}})^2$	X_F	$\bar{X}_F - \bar{\bar{X}}$	$(\bar{X}_F - \bar{\bar{X}})^2$
21			16			12		
21			15			10		
20			15			10		
19			15			10		
19			14			8		

_____ 13. $\Sigma(\overline{X}-\overline{\overline{X}})^2$ _____ 14. n

_____ 15. SS_A _____ 16. df_A

_____ 17. s_A^2

Next, compute SS_A via the raw-score formula on p. 315.

_____ 18. The term in square brackets in Formula 18.2

_____ 19. The other term in the formula, the one subtracted from the first

_____ 20. SS_A as computed from Formula 18.2. Compare with the previous result.

Now complete the analysis:

_____ 21. F_{calc} _____ 22. F_{crit} for $\alpha = .05$

Accept   Reject 23. Decision on null hypothesis stating equality of population means

For a bit of extra insight, finally, compute SS_T first via the definitional formula on p. 313 and then via the raw-score formula on p. 315. To use the definitional formula, fill in the table below, which lists all the scores.

X	$X-\overline{\overline{X}}$	$(X-\overline{\overline{X}})^2$
21		
21		
20		
19		
19		
16		
15		
15		
15		
14		
12		
10		
10		
10		
8		

_____ 24. SS_T from the table, which should equal $SS_W + SS_A$

_____ 25. ΣX^2, where summation is over all scores (this quantity is a part of Formula 18.3)

_____ 26. The other term in Formula 18.3

_____ 27. SS_T from Formula 18.3

_____ 28. What is the critical HSD for Tukey's test at the .05 level of significance?

_____ 29. By the HSD test, which pairs of means are significantly different? Circle those that are:

$\overline{X}_D - \overline{X}_E$ $\overline{X}_D - \overline{X}_F$ $\overline{X}_E - \overline{X}_F$

HOMEWORK FOR CHAPTER 19

In a study concerned with self-awareness, observers watched college students walking past a 25-foot row of reflective windows that functioned like mirrors. An observer stationed inside the building, invisible to the subjects, recorded the duration of time during which a student looked at his or her reflection in the windows. Two other observers independently categorized each student as low, medium, or high in physical attractiveness. Only those students who were categorized identically by the latter two observers were retained as subjects in the study. Observations continued until there were 12 male and 12 female students in each category of attractiveness. The researchers reported the following table; each entry is the mean amount of time in seconds during which the 12 subjects in the given cell gazed at their reflections.

		Sex of Subject	
		Female	Male
Level of Physical Attractiveness	Low	1.0	1.0
	Medium	2.3	2.2
	High	3.5	2.3

One-Way   Two-Factor   1. Did this study employ a one-way design or a two-factor design?

2. What were the treatment variables?

3. What was the response variable (the dependent variable*)?

4. What was the row factor?

_____ 5. How many levels were there in the row factor?

6. Name the levels of the row factor.

7. What was the column factor?

_____ 8. How many levels were there in the column factor?

9. Name the levels of the column factor.

_____ 10. How many cells were there in this study?

* See the footnote on p. 229 of the text.

Dependent   Independent   11. Were the various groups of subjects (there was one group per cell) dependent or independent?

Yes   No   12. Is the factorial analysis of variance described in Ch. 19 appropriate for these data?

_____   13. To two decimal places, what was the mean amount of time spent gazing at the reflection for male subjects?

_____   14. To two decimal places, what was the mean amount of time spent gazing at the reflection for female subjects?

_____   15. To two decimal places, what was the mean amount of time spent gazing at the reflection for the subjects judged to be low in physical attractiveness?

_____   16. To two decimal places, what was the mean amount of time spent gazing at the reflection for the subjects judged to be medium in attractiveness?

_____   17. To two decimal places, what was the mean amount of time spent gazing at the reflection for the subjects judged to be high in attractiveness?

_____   18. In symbols, what is the null hypothesis for a test of the main effect for the row variable?

_____   19. In symbols, what is the null hypothesis for a test of the main effect for the column variable?

_____   20. To two decimal places, what is $\overline{\overline{X}}$?

Yes   No   21. Do the means in the table show that there is an interaction in the samples between sex and level of physical attractiveness?

_____   22. What are the df for testing the main effect of sex of subject?

_____   23. What are the df for testing the main effect of physical attractiveness?

_____   24. What are the df for testing the interaction?

_____   25. What are the within-cells df?

_____   26. What are the total df?

Fill in the following table with the critical values of F. If the df for the denominator of an F ratio are not in the relevant appendix of your text, use the closest df that are in the table.

	F_{crit} at $\alpha = .05$	F_{crit} at $\alpha = .01$
Sex	27.	28.
Attractiveness	29.	30.
Interaction	31.	32.

The researchers reported that the F value for attractiveness was 4.50.

Yes   No   33. Was this result statistically significant at the .05 level?

Yes   No   34. Was this result significant at the .01 level?

According to the researchers, the other effects in the analysis of variance were nonsignificant.

This study was the work of P. J. McDonald & V. C. Eilenfield: "Physical Attractiveness and the Approach/Avoidance of Self-Awareness," *Personality and Social Psychology Bulletin*, 1980, 6, 391-395.

HOMEWORK FOR CHAPTER 20

Consider births that deliver a single child (ignoring twins, triplets, and other multiple births). Assume that the sex of a child is always unambiguous, and that the probability the child will be female is .50. (See the homework for Ch. 9 for a note on this matter.)

Yes No 1. Under these assumptions, is the observation of the sex of a child dichotomous?

_____ 2. Justify your answer to Question 1.

Now use Appendix K to answer the following questions, giving the probabilities or proportions that they request to four decimal places.

_____ 3. What is the probability that seven women who each bear a single child will all have a girl? (Compare your answer with the number you calculated in the homework for Ch. 9.)

_____ 4. What is the probability that 14 women who each bear a single child will all have a girl? (You might think that doubling the number of women would make the event half as likely, but it becomes much less than half as likely.)

_____ 5. What is the probability that seven women who each bear a single child will produce a total of six or more girls?

_____ 6. What is the probability that 14 women who each bear a single child will produce a total of 12 or more girls?

_____ 7. Of the women in the United States who have borne exactly three children, what proportion had no girls?

_____ 8. Of the women in the United States who have borne exactly three children, what proportion had exactly one girl?

_____ 9. Of the women in the United States who have borne exactly three children, what proportion had exactly two girls?

_____ 10. Of the women in the United States who have borne exactly four children, what proportion had over half girls?

_____ 11. Of the women in the United States who have borne exactly six children, what proportion had over half girls?

_____ 12. Of the women in the United States who have borne exactly twelve children, what proportion had over half girls?

In laboratories around the world, researchers are working to discover or create the means to give a couple control over the sex of any child they bear. Let's imagine that an obstetrician reports on a series of 12 cases in which a couple decided to try to have a boy. To assess the claim that the procedure this obstetrician employed has raised the probability of bearing a boy to over .50, we could employ a hypothesis-testing procedure.

_____ 13. What would the null hypothesis state?

_____ 14. Because the procedure might actually lower the probability of getting a boy, it would seem prudent to conduct a test that could discover either that the probability has increased or that it has decreased. State the appropriate alternative hypothesis.

_____ 15. Suppose we measure the outcome of the series of 12 cases as the number of instances in which the baby's sex is male. What outcomes would fall into the region of acceptance if we worked as close as possible to the .05 level of significance?

_____ 16. What outcomes would fall into the region of rejection if we worked as close as we could to the .05 level of significance?

_____ 17. What would the actual level of significance be?

_____ 18. If we work as close as we can to the .01 level of significance, what outcomes would fall into the region of acceptance?

_____ 19. If we work as close as we can to the .01 level of significance, what outcomes would fall into the region of rejection?

_____ 20. What would the actual level of significance be (to four decimal places) in our attempt to work at the .01 level?

Accept   Reject 21. If the researcher reported that 10 of the 12 couples she treated had a boy, what would you conclude about the null hypothesis you were testing—for as close as you can come to the .05 level of significance?

Accept   Reject 22. If the researcher reported that 10 of the 12 couples she treated had a boy, what would you conclude if you were working as close as you could come to the .01 level of significance?

HOMEWORK FOR CHAPTER 21

Want to convince people that you can read their minds? Try this demonstration. Ask a good-sized group of people each to think of a number between six and ten, inclusive. Each person should make his or her choice individually and keep it private. Then request that the group think their numbers "at" you, and announce that you will receive, via telepathy, the number that "comes through" most strongly, which will be the modal choice. Pretend to receive their thoughts, and state with confidence that the "loudest" number is seven. You will have an excellent chance of being correct. Why? The following data supply a clue. They are the results of asking 207 introductory-psychology students to choose a number from six to ten.*

Choice	f_o	f_e	$f_o - f_e$	$(f_o - f_e)^2$	$(f_o - f_e)^2/f_e$
six	24				
seven	112				
eight	33				
nine	25				
ten	13				

Does it appear plausible, in light of these data, that people in our contemporary society make those five possible choices in equal proportions? Test the appropriate hypothesis at the .05 level of significance.

1. State the null hypothesis in words.

2. State the alternative hypothesis in words.

3. Compute χ^2, showing your work in the table above.

    4. $df=$_____   5. $\chi^2_{crit}=$_____   6. Decision on H_0: Accept   Reject

---

* The data were reported by Philip Zimbardo in the instructor's manual for the ninth edition of his text *Psychology & Life*.

Is it plausible that the proportion of people who choose seven is one-half in the population? Compute the appropriate χ^2, showing your work in the table below.

Choice	f_o	f_e	$f_o - f_e$	$(f_o - f_e)^2$	$(f_o - f_e)^2/f_e$
seven					
other					

7. $\chi^2 =$

8. $df=$ _____   9. χ^2_{crit} for $\alpha = .05$: _____   10. Decision on H_0: Accept   Reject

In 1975, three psychologists at Purdue University reported a study in which undergraduate students, 28 men and 34 women, were asked to play an electronic dart game. Each subject was offered a choice between two versions, one in which the score depended on the player's skill and one in which the score depended on luck. The psychologists reported their data in the following table:

		Males	Females
Choice	Luck	6	22
	Skill	22	12

11. State an appropriate null hypothesis in words.

12. State a two-tailed alternative in words.

13. In the table above, write the f_e values next to their f_o counterparts. Give the f_e values to three decimal places.

14. χ^2 to two decimal places. Show your work below.

15. $df=$ _____   16. χ^2_{crit} for $\alpha = .05$: _____   17. Decision on H_0: Accept   Reject

This study was the work of Kay Deaux, Leonard White, and Elizabeth Farris: "Skill versus Luck: Field and Laboratory Studies of Male and Female Preferences," *Journal of Personality and Social Psychology*, 1975, *32*, 629-636.

HOMEWORK FOR CHAPTER 22

In a replication of the study described in the homework for Ch. 15, a female experimenter directed a stare or just a glance at a pedestrian waiting for a traffic light to change from red to green. A second experimenter standing across the street timed the subject as she or he crossed the intersection after the light changed. Crossing times were recorded to the nearest half second, and the following (real) data resulted:

Stare:	6 7 7.5 7.5 8 8 8 8	9	12.5
Glance:		8.5 8.5 9 9.5 10 10 10 10.5 11 11	

Test the difference between the two samples via the Mann-Whitney procedure, using a two-tailed alternative and the .05 level of significance. Call the stare condition X, and show your work in the table above.

_____ 1. ΣR_X

_____ 2. Critical value(s)

Accept   Reject 3. Decision on H_0

_____ 4. If you were doing a one-tailed test at the .05 level of significance and expected to find crossing times faster in the stare condition, what pair of values in Appendix M would be relevant?

Accept   Reject 5. What would your decision on H_0 be?

_____ 6. If you were doing a one-tailed test at the .01 level of significance and expected to find crossing times slower in the stare condition, what pair of values in Appendix M would be relevant?

Accept   Reject 7. What would your decision on H_0 be?

In the appendix to this *Guide* is a bivariate distribution of scores on statistics tests. One column gives percentages correct for the first exam in a statistics course for 10 students, and the other column gives percentages correct over the entire semester for those same students.

Independent   Dependent 8. Are the two samples of scores independent or dependent?

Conduct a sign test of the location of the two parent distributions. Assign a + to each student whose percentage correct on the first exam exceeded his or her percentage correct over the semester; assign a − to each student whose percentage correct over the semester was higher.

_____ 9. How many pluses are there?

_____ 10. How many minuses are there?

11. State in words an appropriate H_0.

12. State in words the appropriate nondirectional alternative hypothesis.

Calculate χ^2, showing your work in the table below.

	f_o	f_e	$f_o - f_e$	$(f_o - f_e)^2$	$(f_o - f_e)^2 / f_e$
+					
−					

13. $\chi^2 =$

_____ 14. *df*

_____ 15. Critical value(s) of χ^2 at the .01 level of significance for a two-tailed test

Accept Reject 16. Decision on H_0

There was an eleventh person in the class, as you may recall from the homework for Ch. 8. This student scored 69% on the first exam and 86% over the entire semester. Redo the sign test with these data included.

_____ 17. Number of pluses.

_____ 18. Number of minuses.

Calculate χ^2, showing your work in the table below.

	f_o	f_e	$f_o - f_e$	$(f_o - f_e)^2$	$(f_o - f_e)^2 / f_e$
+					
−					

19. $\chi^2 =$

_____ 20. *df*

_____ 21. Critical value(s) of χ^2 at the .005 level of significance for a two-tailed test.

Accept Reject 22. Decision on H_0 (at the .005 level)